Intellectual Property Rights at a Glance

The Authors

Phundan Singh [b. 1946] hails from a reputed agricultural family of Western Uttar Pradesh [Village-Sakauti, Tehsil- Mawana, District- Meerut]. He graduated in Agriculture from Agra University, Agra and obtained his Master and Doctorate degrees from Kanpur University. He has throughout a brilliant academic record. His area of specialization is Plant Breeding and Genetics. He was selected as Scientist of first competitive examination conducted by ICAR through ASRB.

Dr. P. Singh has total experience of more than 40 years in the field of Plant Breeding and Genetics. He has served the Central Institute for Cotton Research, Nagpur for about 32 years in various capacities such as Scientist, Senior Scientist, Principal Scientist, Head, Division of Crop Improvement and Director. He has about 300 publications to his credit. He has participated and presented 80 research papers in various National and International Seminars, Symposia and Conferences. He has authored 52 books mostly published by Kalyani Publishers, New Delhi, 18 Scientific bulletins and contributed 14 chapters in various reputed books.

Dr. P. Singh is a member of several Scientific Societies, referee of different journals, Expert member of Selection Committees, M. Sc. and Ph.D. Examiner and paper setter in different Agricultural Universities. The CICR was recognized as the referral Laboratory for detection of Bt. gene and received the best Annual report Award in 2003 when he was the Director. He is a widely travelled Scientist. He has visited several countries such as Russia, Belarus, Ukraine, Crimea, Uzbekistan, Canada, USA and England.

Engineer Rajeev Singh graduated in Mechanical Engineering from Visvesvaraya National Institute of Technology, Nagpur in 1st Division. Based on campus interview, he was selected as Assistant Manager by Reliance Industries Limited (RIL). Later on he was elevated to the post of Manager, Senior Manager, General Manager and Senior General Manager. Presently, he is working as Senior General Manager with Reliance Industries Limited. He has throughout a brilliant academic record. He has more than 27 year managerial experience. He has authored one book entitled IPR and Plant Breeders' Rights.

He has visited two countries, *viz.* United States of America and Republic of China for official work. He obtained Gold Medal in shooting competition at Surat organized by Surat City Rifle Club. He obtained Certificate in Rifle Shooting from Swarnim Gujarat Sports University and got Gold Medal topping the list of successful candidates. His hobbies are running, cycling, shooting, *etc.*

Intellectual Property Rights at a Glance

— *Authors* —

Phundan Singh

Former Director
Central Institute for Cotton Research,
Nagpur – 440 010
Maharashtra, India

and

Er. Rajeev Singh

Senior General Manager
Reliance Industries Limited,
Surat (Gujarat)

2018

Daya Publishing House®

A Division of

Astral International Pvt. Ltd.

New Delhi – 110 002

Cataloging in Publication Data--DK
Courtesy: D.K. Agencies (P) Ltd. <docinfo@dkagencies.com>

Singh, Phundan, 1946- **author.**
Intellectual property rights at a glance / authors, Phundan Singh and Er. Rajeev Singh.
pages cm
Includes bibliographical references and index.
ISBN 9789387057685 (International Edition)

1. Intellectual property--India. 2. Intellectual property. I. Singh, Rajeev (Corporate manager), author. II. Title.

LCC KNS1100.S56 2018 | DDC 346.54048 23

Published by : **Daya Publishing House®**
A Division of
Astral International Pvt. Ltd.
– ISO 9001:2015 Certified Company –
4736/23, Ansari Road, Darya Ganj
New Delhi-110 002
Ph. 011-43549197, 23278134
E-mail: info@astralint.com
Website: www.astralint.com

Dedicated to Granddaughters of Senior Author

Miss Rishika Singh

and

Miss Muskaan Singh

Preface

The subject of Intellectual Property Rights is gaining increasing importance these days. There are several books on this subject both of foreign and Indian origin. However, almost all these books deal with the subject in a descriptive manner. There is hardly any book which treats the subject in short questions and answers form. The prime objective of writing this book is to fill up this gap and present the subject in questions and answers form. This book has been designed for competitive examinations such as JRF, SRF, NET, SET, bank officers, agriculture officers, *etc.*

This book contains 20 chapters which have been divided into two sections. Section I deals with Primary Rights comprising 12 chapters [Chapter 1 to Chapter 12]. And Section II deals with *Sui-generis* Rights consisting of 8 Chapters [Chapter 13 to Chapter 20]. The information contained in this book has been gathered from various published sources and Internet websites. Attempts have been made to provide latest information even then some valuable information might have been missed.

The summary of Intellectual Property Rights, addresses of Copyrights, Patent and trademarks Offices in India and lists of domain names of some 50 countries have been appended. The Glossary of important technical terms used in relation to Intellectual Property Rights has been provided at the end for ready reference. Important References have been listed at the end which would be useful in getting further details.

It requires long hours to complete job like this. Completion of this manuscript took more than three years. I could not spare adequate time for my family members during writing of this book for which I felt guilty. The cooperation extended by my wife Mrs. Jaswanti Singh during preparation of this Volume is highly appreciable

Hope this Volume would be useful to the students, researchers, Plan Breeders, Design developers, artists, musicians, private seed companies, *etc.* Constructive suggestions of readers are welcome for further improvement of this book.

Phundan Singh

Rajeev Singh

Contents

Section I

Primary Rights

Chapter 1

Outlines of Intellectual Property Rights

A. PRIMARY RIGHTS

1. What do you mean by a property?

The property refers to wealth or valuable things earned by a person. It is estimated in terms of land, house, garden, industries, animals, gold, silver, diamond, money *etc.* accumulated by a person.

2. What is the need to earn property?

Each person wants to earn property for four main purposes as given below:

(i) To get a better or higher status in the society, because status of a person is measured in terms of property accumulated by him or her. On the basis of property, the person is called as rich, middle class or poor.

(ii) To lead a better quality of life in the society.

(iii) To get a name and fame in the society. In the society, rich people get more respect than others.

(iv) For better financial security of life for himself and for his children.

3. What are different types of property?

The various valuable belongings of a person are broadly classified into two groups, *viz.*, (i). immovable property, and (ii). movable property.

4. What is Immovable Property?

It refers to fixed types of properties which can not move from one place to other. Such properties include land, buildings [house, flat, bungalow, farm house] and gardens. Such properties are protected by laws of land in each country.

5. Define Movable Property

It refers to the property which can be easily shifted from one place to other. Such property includes animals, farm machines, furniture, fixtures, gold, silver, diamond, and money. Such properties are also protected by public laws.

6. What is Intellectual Property?

There is third type of property what we call intellectual property. The product/ process/idea which is outcome of the brain of a person and can be used on commercial scale for benefit of human kind is called intellectual property. In other words, Intellectual property refers to creations of the mind: inventions, literary and artistic works, and symbols, names, images, and designs used on commercial scale.

7. Describe main features of Intellectual Property

The main features of intellectual property are given below:

1. It is measured in terms of new ideas, processes, products, inventions and innovations developed by a person.
2. It requires lot of intellectual inputs in terms of thinking, planning and fine tuning of new ideas/products/processes, *etc.*
3. It requires considerable amount of funds and other resources to develop new products/processes.
4. The main problem with intellectual property is that it can be copied, reproduced and used by others resulting in loss of inventor. Hence protection of intellectual property is essential so that the inventor can derive maximum benefits from his invention.

8. What do you mean by Intellectual Property Rights?

The rights to intangible property that is the product of the human intellect are referred to as intellectual property rights. Intellectual property may be protected by copyright, trademark or patent. The holder of intellectual property rights is usually the person or persons who developed the product or the organization that funded it.

9. What are different forms of Intellectual Property Rights?

The Intellectual Property Rights are broadly divided into two groups, *viz.*, Primary Rights and *Sui-generis* Rights. These are briefly discussed below:

10. What do you mean by Primary Rights?

Primary rights include,Copyrights, Patents, Trademarks, Trade Secret, Trade names, Domain names, Industrial Designs,Geographical Indications, *etc.*

11. Define Sui-generis Rights

Sui-generis refers to things of their own kind or things with unique characteristics. Such rights include Database Rights, Mask work, Plant Breeders Rights, Traditional Knowledge, Moral Rights and Supplementary Protection Certificate.

12. Give a list of different types of Intellectual Property Rights

A list of different types of Intellectual Property Rights is presented in the following Table 1.1.

TABLE 1.1: Forms of Intellectual Property Rights/Laws

Sl.No.	*Primary Rights*	*Sui-Generis Rights*
1	Copyrights	Database Rights
2	Patents	Mask work
3	Trademarks	Plant Breeders' Rights
4	Industrial Design Rights	Farmers' Rights
5	Utility Models	Moral Rights
6	Geographical Indication	Supplementary Protection Certificate
7	Trade Secret	Indigenous Intellectual Property
8	Related Rights	
9	Trade Names	
10	Domain Names	

13. What do you mean by Copyrights?

Copyright refers to a document which grants exclusive right to the author/ creator to publish and sell literary or musical or artistic work. It provides exclusive rights to the owner to reproduce, sell and distribute a work, prepare derivative works and display the work publicly. It includes right of an author, artist, publisher, *etc.* to retain ownership of works and to produce or contract others to produce copies. Thus, it refers to the legal right granted to an author, composer, playwright, publisher, or distributor for exclusive publication, production, sale, or distribution of a literary, musical, dramatic, or artistic work.

14. What are the main features of copyrights?

Main features of copyrights are briefly presented below:

(i) It provides protection for a specific duration.

(ii) It is applicable in all countries.

(iii) The copyright holder has the rights to authorize others to use the protected work.

(iv) It is applicable to books, movies, music, paintings, photographs and software.

(v) The information cannot be reproduced as such without written permission. However, the information or idea can be used by any one.

15. What is a Patent ?

Patent has been defined in various ways by different authors.Patent refers to a document granting an inventor sole rights to an invention. It is an official document which grants sole rights to the inventor for manufacturing and marketing his product/process/invention to derive benefits. In other words, patent is a set of exclusive rights granted by a State to a patentee (the inventor or assignee) for a fixed period of time.

16. Explain briefly main points of a patent.

The main points related to a patent are briefly presented as follows.

(i) It is a legal document issued by a federal government that grants exclusive rights for the production, sale and profit from the invention of a product or process for a specific period of time. Patents also grant the right to prevent others from copying the invention.

(ii) It is an official license granted by the Patent Office to issue exclusive right to an individual or business for production or sale of a specific invention.

(iii) It is a certificate issued by a government to an inventor, assuring the sole right to make, use, and sell the invention for a certain period of time. Many audio and video technologies are covered by patents.

(iv) It is a document that allows the patent owner to prevent others from making, using or selling the invention protected by the patent.

17. What is a trademark?

A trademark or trade mark is a distinctive sign of some kind which is used by an individual, business organization or other legal entity to uniquely identify the source of its products and/or services to consumers, and to distinguish its products or services from those of other entities.

18. What are main features of a trademark?

The main features of trademarks are briefly presented below:

(i) A trademark can be a word, name, symbol, device or mark which is used to identify and distinguish the goods or services of one company from goods or services of another.

(ii) Trademark is used to identify its product and to distinguish them from others. It is the name of a product made by a particular person or company.

(iii) The period of protection for a trademark varies, but can generally be renewed indefinitely.

19. What do you mean by Industrial design ?

An industrial design - or simply a design - is the ornamental or aesthetic aspect of an article produced by industry or handicraft; registration and renewals provide protection for, in most cases, up to 15 years.

20. Describe main points related to an industrial design.

The main points about industrial designs are given below:

(i) Industrial design is an applied art whereby the aesthetics and usability of products may be improved for marketability and production. Industrial

Designers often utilize 3D. Industrial Design is the field of developing physical solutions to meet a particular need. These physical solutions might include products.

(ii) Industrial design refers to any original shape, picture, or some combination applied to a useful article of manufacture. In other words, it refers to the design of the mass-produced products of our everyday environment, from sinks and furniture to computers.

(iii) Industrial design refers to the professional service of creating and developing concepts and specifications that optimize the function, value, and appearance of products and systems for the mutual benefit of both user and manufacturer [Industrial Design Society of America].

21. What are utility models?

A utility model is an exclusive right granted for an invention, which allows the right holder to prevent others from commercially using the protected invention, without his authorization, for a limited period of time. In its basic definition, which may vary from one country (where such protection is available) to another, a utility model is similar to a patent. In fact, utility models are also referred to as "petty patents" or "innovation patents, minor patent and small patent."

22. Describe different points related to utility models.

The important points about utility model are listed below:

(i) Utility model is an intellectual property right to protect inventions.

(ii) Utility models are also known as petty patents, petty inventions, petty innovations, utility innovations, minor patents and small patent.

(iii) The rights conferred by utility models are similar to those granted by patent laws but has a shorter term.

(iv) They are registered with National Patent Office.

(v) They are more suited to incremental inventions.

23. What do you mean by geographical indication?

Geographical Indications [GI] is the name of a region, a specific place or, in exceptional cases a country, used to describe an agricultural, natural or manufactured goods product or a food stuff. In other words, a geographical indication is a sign used on goods that have a specific geographical origin and often possess qualities or a reputation that are due to that place of origin. In other words, a geographical indication is a name or sign used on certain products or which corresponds to a specific geographical location or origin (*e.g.* a town, region, or country)

24. Write short notes on geographical indication

Geographical Indication may be an agricultural, natural or manufactured goods or product. So far geographical indications have been registered in some advanced countries because they have developed the system of protecting GI. In the past, wines,spirits, cheeses, tobacco, which account for 88 per cent, have been registered as GI. Wines and spirits account for almost 71 per cent of all registrations. In addition to wines and spirits, there is need for protection of agricultural products as GI. For this there is need to establish a multilateral system of identification and registration of GI.

25. Define Trade Secret

Trade secret may be a formula or process or device used in business that is not published or divulged which gives an advantage over competitors. It is, known only to its owner, that is used in producing something of commercial value.

26. Describe main points related to trade secret.

The main points about trade secret are briefly presented as follows:

(i) There is no specific period for trade secret. It may continue life long or for generations together.

(ii) There is no need of registration for trade secret.

(iii) Iit does not provide opportunity to others for improvement of innovation.

(iv) It is not applicable to books, equipments, plant varieties, designs which are openly used.

27. What are Related Rights?

Related rights is a term in copyright law, used in opposition to the term "authors' rights". The term neighbouring rights is exactly equivalent, and a more literal translation of the original French *droits voisins*. Related rights in civil law are rights which are similar to authors' right but which are not connected with the actual author of the work. Both authors' rights and related rights are copyrights in the sense of English or U.S. law.

28. Explain main points about related rights.

The main points about related rights are given below:

1. Related rights are similar to copyrights.
2. Related rights come under primary property rights.
3. Related rights are also known as neighouring rights.
4. Related rights are closer to copyrights but are not covered by the Berne Convention.
5. Related Rights are covered by Rome Convention, 1961.

29. What do you mean by trade name?

A trade name, also known as a trading name or a business name, is the name which a business trades under for commercial purposes, although its registered, legal name, used for contracts and other formal situations, may be another. Pharmaceuticals also have trade names (*e.g.* "Aspirin"), often dissimilar to their chemical names ("acetylsalicylic acid"). Trading names are sometimes registered as trademarks or are regarded as brands. A trade name is the name under which a product is commercially known. The commercial name by which a chemical is known is called trade name. One chemical may have a variety of trade names depending on the manufacturers or distributors involved.

30. Describe main points about Trade Names.

The main points related to trade name are given below:

(i) It is also known as business name, trading name, assumed name, brand name and corporate name.

(ii) It may or may not be registered

(iii) It distinguishes a particular business from others.

(iv) It can be exclusive or non-exclusive.

(v) It is generally used on letter heads and bank accounts.

(vi) It is the name given to a particular substance by each company that manufactures it.

(vii) It is the name under which a company conducts its business.

(viii) It is s the business name of the person or organization making and/or selling a particular product. "General Motors, Inc" is the "trade name" of the company

31. What do you mean by Domain name?

Domain name refers to the unique name that identifies an Internet site. Domain names have two or more parts separated by dots. For example www.ask-edi.com or www.bcbsks.com The term domain name has multiple related meanings:

(i) It identifies a web site on the internet (*e.g.*, company.com); also referred to as a URL [Uniform Resource Locator].

(ii) It allows reference to Internet sites without knowing the true numerical address.

(iii) It refers to the official name of a computer connected to the Internet (*e.g.*, "www.microsoft.com").

(iv) It is the unique name of a computer on the Internet that distinguishes it from other systems around the world.

(v) They are sometimes colloquially (and incorrectly) referred to by marketers as "web addresses".

32. What are the Advantages of Intellectual Property Rights ?

Intellectual Property Rights refer to the legal rights provided to an inventor to derive economic benefits from his invention/innovation. The main advantages of IPR are as follows:

(i) It promotes healthy competition for invention/innovation among the intellectuals.

(ii) It helps in improving the quality of the product.

(iii) It makes available new ideas/technologies to different countries.

(iv) It leads to faster development of industries/organizations engaged in research and development work.

33. What the disadvantages of Intellectual Property Rights ?

There are some disadvantages of Intellectual Property Rights which are briefly presented below:

(i) The procedure of registration, particularly of patents, is very lengthy.

(ii) It involves lot of money transaction in registration, renewal and licensing.

(iii) It invites lots of court cases due to infringements.

(iv) It may lead to monopoly of right holders, *etc.*

B. *SUI-GENERIS* RIGHTS

34. Define *sui-generis* rights.

Sui-generis is a Latin expression, literally meaning of its own kind or unique in its characteristics. In law, it is term of art used to identify a legal classification that exists independently of other categorizations because of its uniqueness or due to the specific creation of an entitlement or obligation. In Intellectual Property there are rights which are known as being sui generic to owners of a small class of works. Such Intellectual Property Rights include (i) database Rights, (ii) mask work, (iii) plant breeders' rights, (iv) farmers' rights, (v) moral rights, (vi) supplementary protection certificate, and (vii) indigenous intellectual property.

35. What do you mean by Database rights?

Database rights are a form of exclusive right introduced by European Union Law in 1996 in those countries which follow EU Law. A 'database right' is an intellectual property right given to a computer database. The main points about database rights are as follows:

(i) In most countries database are covered by copyright law to some degree.

(ii) The database rights were introduced by European Union Law to provide greater protection to collection of information.

(iii) Database right is covered by sui generic rights.

(iv) **Duration.** Database rights last for 15 years, but can be extended if the database is updated.

(v) **Protection**. Database right prevents copying of substantial parts of database. The protection is over the information, not over the form of expression.

36. What are Mask Work Rights?

A mask work is a two or three-dimensional layout of an integrated circuit (IC or "chip"), *i.e.* the arrangement on a chip of semiconductor devices such as transistors and passive electronic components such as resistors and interconnections. Main points related to mask work are listed as under:

(i) Mask work can not be protected under copyright law or patent law.

(ii) The mask work exclusive rights were first granted in USA by Semiconductor Chip Protection Act of 1984. Such legislations also exist in Canada and Australia.

(iii) Duration. Rights for semiconductor mask work lasts only for two years (if unregistered) or 10 years (if registered).

(iv) The exclusive rights in mask work are somewhat like those of copyright: the right to reproduce the mask work or distribute an IC made using mask work.

(v) Reproduction for reverse engineering of a mask work is permitted by law.

(vi) Mask work rights exist when they are created, even without registration.

(vii) Protection under mask work cannot be provided to a mask that is not original.

37. What do you mean by plant breeders' rights?

Plant breeders' rights, also known as plant variety rights (PVR), are intellectual property rights granted to the breeder of a new variety of plant. Plant Breeders Rights are granted to novel plant varieties that are distinctive, uniform, and stable (*e.g.*, cultivars bred true-to-type for desired traits). The legal protection of a new plant variety is granted to the breeder or his successor. The effect of PBR is that prior authorization is required before the material can be used for commercial purposes.

38. What are Advantages of PBR?

There are several advantages of plant breeders' rights which are listed below:

(i) Breeders get benefit of their variety

(ii) PBR help in faster development of seed industry.

(iii) PBR lead to improvement in quality because of competition

(iv) PBR are useful in procurement of good material on payment basis.

(v) PBR help in enrichment of genetic resources.

39. What are Disadvantages of PBR?

There are several advantages of plant breeders' rights which are listed below:

(i) It will promote monopoly.

(ii) It will encourage unhealthy practices.

(iii) It may lead to increase in prices.

(iv) There will be reduction in genetic variability.

(v) There will be compulsion to purchase frsh seed every year.

40. What are Farmers' Rights?

Farmers' rights refer to the rights arising from the past, present and future contributions of farmers in conserving, improving and making available plant or animal genetic resources, particularly those in the centres of origin/diversity. In other words, the legal rights provided to farmers to save, use, sow, replant, exchange, share or sell his farm produce including seed of a variety protected under Plant Variety Protection Act refer to Farmers Rights. The purpose of these rights is to "ensure full benefits to farmers and support the continuation of their contribution.s

The FAO Conference held in Rome from 11-29 November, 1989 endorsed the concept of Farmers Rights with a view to:

(i) Ensuring global recognition of the need for conservation and the availability of sufficient funds for these purposes;

(ii) Assisting farmers and farming communities throughout the world, especially those in areas of original diversity of plant genetic resources, in the protection and conservation of their PGR and of the natural biosphere; and,

(iii) Allowing the full participation of farmers, their communities and countries in the benefits derived, at present and in the future, from the improved use of PGR.

41. What are moral rights?

Moral rights are rights of creators of copyrighted works generally recognized in civil law jurisdictions and first recognized in France and Germany, before they were included in the Berne Convention for the Protection of Literary and Artistic Works in 1928. Moral rights are a special extension to copyright that gives the originator of a copyrighted work rights over its use. The originator can specify moral conditions on the use or exploitation of their creation, even when the rights are licensed or sold to others'.

42 Define Supplementary protection certificate.

This term is used in European Countries. This is an extension of the term given to pharmaceutical or plant protection patent. A supplementary protection certificate (SPC) is a *sui-generis*, patent-like, intellectual property right. This type of right is available for medicinal products, such as drugs, and plant protection products, such as insecticides, and herbicides.

43. What do you mean by Traditional Knowledge?

The Indigenous Knowledge also known as Traditional Knowledge [TK] refers to the knowledge, innovations and practices of indigenous people and local communities. In other words, it is the: knowledge gained through tradition from ancestors about plants and animals, medicines, agricultural practices and other matters.

Chapter 2

Copyright Law

1. What do you mean by Copyrights?

Copyright refers to a document which grants exclusive right to the author/creator to publish and sell literary or musical or artistic work. It provides exclusive rights to the owner to reproduce, sell and distribute a work, prepare derivative works and display the work publicly. It includes right of an author, artist, publisher, *etc.* to retain ownership of works and to produce or contract others to produce copies. Thus, it refers to the legal right granted to an author, composer, playwright, publisher, or distributor for exclusive publication, production, sale, or distribution of a literary, musical, dramatic, or artistic work.

2. What is the duration of the protection of a copyright?

Copyright provides protection for a specific duration. In India, the duration of copyright is 60 years after the death of the Author. For a work for hire or for an anonymous or pseudonymous work, the copyright lasts for 85 years after publication or 110 years after creation whichever is earlier.

3. Is registration compulsory for copyright work ?

The registration is not compulsory for copyright. In other words, it is an unregistered right. Since there is no registration for copyright, there is no need of filling application and there is no fee for copyright. However, it is advisable to register the copyright with the copyright office. It will provide additional legal rights to the copyright holder.

4. Can we transfer copyright to other person ?

The copyright holder has the rights to authorize other persons to use the protected work. In other words, the author may transfer the copyright to any other party if he so desires.

5. Who is the owner of copyrights ?

The author or creator of the new work is the owner of the copyrights. It is inherited as ordinary property. The information cannot be reproduced as such without written permission of the author or creator. However, the information or idea can be used by any one for creating the new work. The owner of a copyright can exclude others from reproducing, adapting, publicly distributing, publicly displaying or performing the copyrighted work.

6. When does the copyright come into force?

The copyright comes into effect immediately after release of the matter covered by copyright such as book, movie, sound recording, *etc.*

7. What are the Rights Protected by copyrights ?

The purpose of copyright law is to encourage creative work by granting a temporary monopoly in an author's original creation. It controls the right of:

(i) Reproduction of work *i.e.* copying,
(ii) Creation of derivative work,
(iii) Distribution of work,
(iv) Performance of the work,
(v) Display of the work, and
(vi) Digital transmission of performance.

The law of copyright protects the first two rights in both private and public contexts, whereas an author can only restrict the last four rights in the public sphere.

8. What do you mean by copyright infringement?

The copyright Infringement refers to an unauthorized use of material protected by copyright. Infringement of the copyrighted material is possible. Because it an unregistered right, the official action cannot be taken against infringement of unregistered copyrights. The following acts by a person other than the author amount to infringement:

(i) Reproducing or copying of the work,
(ii) Creating a new work derived from the original work such translating the work into a new language and transferring the work into a new medium of expression,
(iii) Sell or give away the work or a copy of the work. Before transfer of copyright to that person, and
(iv) Performing or displaying the work in public without permission from the copyright owner.

9. What are the Defenses against Infringement of copyrights ?

The primary defense to copyright infringement is the fair use. The fair use is described as the privilege to use the copyrighted material in a reasonable manner without the consent of the owner. However use should not have adverse effect on the marketing of the original author. The following activities are considered as fair uses.

(i) Distributing copies of a section of an article in class for educational purposes,

(ii) Providing a quotation in a book review,and

(iii) Imitating a work for the purpose of parody or social commentary.

10. What are the Requirements for registration of copyright work ?

There are two basic requirements for protection of work under copyright, *viz.*, (1) originality and (2) fixed expression. These are explained as follows:

(1) Originality: The first requirement is originality. Originality means that the work should be independently creared by the author.For originality the work need not necessarily have novelty,artistic merit,truth or lawful contents. For example a replica of painting in the public domain may not be novel, but is copyrightable. Similarly, a false biography is copyrightable.

(2) Fixed Medium of Expression: The work should be in fixed medium of expression. For example if someone performs in a local park it is not copyrightable. But if the performance is recorded in a film, then the film is copyrightable. Single words, short phrases and titles are not protected under copyright, because these are not considered "original works of authorship" under the copyright Act.

11. What are the matters Covered by copyrights?

A wide variety of work is protected under copyright law. For protection under copyright the work should have originality and wide utility. The list of items protected under copyright law is presented below:

(i) **Literary Work:** It includes books [novels, instruction manuals, *etc.*], computer programs, lyrics for songs, articles in news papers and some types of database.

(ii) **Dramatic Work:** It includes works of dance or mime.

(iii) **Artistic Work:** It includes graphics and pictures such as paintings, engravings, photographs, sculptures, collages, works of architecture, technical drawings, diagrams, maps, logos, *etc.*

(iv) **Sound Recordings:** It includes recordings on tape and compact disc. It also includes recordings of other copyright work such as musical or literary.

(v) Musical and choreographic work.

(vi) Audiovisual Work; Such as Movies andvideos; and

(vii) Broad casts.

(viii) **Work for Hire:** A work which is created by an employee in the regular course of his employment, is considered a **work for hire** and the employer, not the employee, is considered the author of the work for copyright purposes.

In brief the copyrights are applicable to books, movies, music, paintings, photographs, sound recordings and software.

12. What are the differences between copyrights and patents?

There are several differences between copyrights and patents which have been presented in Table 2.1

TABLE 2.1: Comparison of Copyrights and Patents

Sl.No.	*Particulars*	*Copyrights*	*Patents*
1	Requirement	Originality and wide utility	New invention
2	Period	70-120 years	15-20 years
3	Registration	Not compulsory	Compulsory
4	Validity	All countries	In original country
5	Fee	No Fee	Fees is charged
6	Legal rights	No legal Rights	There are legal Rights
7	Transfer	Possible	Admisible
8	Items covered	Literary, music, movie etc	Products/process/idea
9	Enforcement	Immediately after release of work	After Registration
10	Control	By Author/ creator	By inventor
11	Infringement	Very common	Possible

13. What are the differences between copyrights and trademarks?

There are several differences between copyrights and trademarks which have been presented in Table 2.2.

TABLE 2. 2: Comparison of Copyrights and Trademarks

Sl.No.	*Particulars*	*Copyrights*	*Trademarks*
1	Requirement	Originality and wide utility	New & distinctive
2	Period	6-110 years	Renewable indefinitely
3	Registration	Not compulsory	Is compulsory
4	Validity	All countries	All Countries
5	Fee	No Fee	Fee is charged
6	Legal rights	No legal Rights	Legal rights are there
7	Transfer	Possible	Possible
8	Items covered	Literary, music, movie, *etc.*	Names, images & short phrases
9	Enforcement	After release of work	After registration
10	Control	By Author/ creator	By owner
11	Infringement	Very common	Possible

14. What are the Advantages of Copyright?

There are several advantages of copyrights which are listed below:

(i) Registration is not compulsory.

(ii) It protects original work.

(iii) There is no fee for copyright protection.

(iv) The protection is for a very long period.

(v) It is valid in all countries.

15. What are the Disadvantages of copyrights?

There are some disadvantages of copyrights which are listed below:

(i) Copyrights do not protect ideas, procedures, processes, systems, methods of operation, concepts, principles or discoveries.

(ii) It does provide legal rights without registration.

(iii) It protects only against copying.

(iv) Do not protect historical facts.

Chapter 3

Rights Related to Patent

1. What is a Patent?

Patent is a legal document issued by a government that grants exclusive rights for the production, sale and profit from the invention of a product or process for a specific period of time. Patents also grant the right to prevent others from copying the invention.

2. What do you mean by full term patent?

Patents have been classified in various ways. On the basis of duration, patents are of two types, *viz.* (1) full-term patents and (2) short-term patents. The maximum period for grant of a patent is 20 years. Those patents that are granted for a period of 20 years are referred to as full-term patents. For the grant of full-term patent, the applicant must provide evidence of the novelty of invention. Thus full-term patents have two main features. They are granted for full-term of 20 years. It requires an evidence of the novelty of invention.

3. What is a short term patent?

Those patents that are granted for less than 20 years are called short-term patents. Such patents are granted for smaller inventors. There are two main features of such patent. They are granted maximum for a period of 10 years. and evidence of the novelty of invention is not required.

4. What is a utility patent?

The patents are also classified as (i) utility patents, (ii) design patents, and (iii) plant patents. A type of patent that is granted to inventions that perform useful functions is referred to as utility patent. This is also known as patent for invention. Main feature of utility patent are listed below:

(i) Such patents are issued for the invention of new and useful process, machine, manufacture or composition of matter and their improvement.

(ii) It is granted for full-term of 20 years.

(iii) About 90 per cent of the patents are utility patents.

5. What is a design patent ?

A type of patent that is granted to visual ornamental characteristics such as shape and surface of an article of manufacture is called design patent. It is also called as picture patent. A design for surface ornamentation is inseparable from the article to which it is applied and cannot exist alone. It is definite pattern of surface ornamentation, applied to an article of manufacture. Main features of utility patent are listed below:

(i) This is granted for a period of 14 years.

(ii) It is also known as picture patent.

(iii) Design patents do not protect an idea or an invention. They only protect ornamental design of exactly what is pictured.

(iv) They are very weak and appropriate in limited circumstances.

6. What is a plant Patent ?

A type of patent that is granted to new or newly developed, distinct plants reproducing asexually. Main features of plant patents are listed below:

(i) These patents are of recent origin.

(ii) They are useful in protecting novel plants or new plants.

(iii) Protected for a period of 5 years.

7. What is the duration of protection for a patent ?

A patent is protected for a specific duration. The period of protection varies from 15 to 20 years. In India it is 7-14 years. The long term patents are protected for 20 years in many countries and short term patents for 14 years.

8. Is registration compulsory for a patent ?

The registration is compulsory for the patent. In other words, patent is a registered right. An application is submitted in prescribed form to the patent office for the registration of the paten and some fee is also paid for registration.

9. What is the validity of a patent?

The patent is valid in the country where it has been granted. In future the patent granted by WTO will be valid in all member countries.

10. Can we transfer a patent?

The inventor has the power to transfer the right of patent to any other person if he so desires. In other words, the inventor may transfer ownership to any one [including a corporation]. Employee often assign the rights of their invention to their employers as part of their employment contracts.

11. What are the matters that can be patented?

The patentable subject matter include five main items, *viz.*, (1) processes, (2) machines (3) articles of manufacture, (4) business methods and (5) software related ideas. Besides, new algorithms, data structure, new applications, business procedure and new technology can be patented. The same software may qualify for both copyright and patent protection, if it satisfies requirements for both of these.

Processes include business methods, most software, medical techniques, sports techniques and the like. Machines include devices and apparatuses. Articles of manufacture include mechanical devices, electrical/electronic devices and composition of matter such as chemicals, medicines, DNA, RNA, *etc.*

In brief, patent is applicable to new process, products, invention, device and innovations which have commercial or industrial applications.

12. What are non-patentable Matters?

The following matters cannot be patented.

(i) Ideas, mental processes, laws of nature and physical phenomena. For example one cannot patent a new mineral, insect or plant discovered or found in the wild.

(ii) Literary, dramatic, musical and artistic work. These are subject to copyright protection.

(iii) Inventions that are not useful or that are offensive to public morality.

13. Who has control over the patent ?

The inventor of the process, product or invention has control over the patent. Patents provide exclusive rights to the inventor for the production, sale and profit from his invention or a product or process for a specific period of time.

14. What is the procedure of filing a patent ?

An inventor cannot patent an inventions if he or she discloses it to the public more than one year before filing application for patent prosecution. Predisclosure is known as on sale which bar to patentability. The legal process of obtaining a patent from a patent office is called patent prosecution. An application for a patent is prepared by a professional patent agent or patent attorney and filed with a patent office where it is examined by a patent examiner for patentability.Patent protection in multiple countries requires filing of separate patent applications in each country. The patent prosecution process involves following steps.

(i) Filing a duly filled up patent application by inventor or applicant.

(ii) Establishing of a prior art search report by the patent office.

(iii) Publication within 18 months from earliest claimed filing date.

(iv) Review by the examiner or the examining Division, including communication with applicant to modify the claim language, if needed.

(v) Grant of the patent [if the patentability criteria are met] and publication of the issued patent.

(vi) Opposition period, during which anybody (*e.g.* other companies) can challenge the patent. In several countries, oppositions can be filed before the grant of the patent.

Publicly disclosed inventions including patented inventions are known as prior art. If prior art is not disclosed by the applicant in his patent application the patent can be invalidated for hiding information.

15. What are the criteria for examination of a patent?

The following criteria are taken into account in the examination process of the patent application submitted by the inventor or applicant.

(i) Verifying that claims are for a patentable subject matter.

(ii) Ensuring unity of invention: there should separate application for each invention.

(iii) Ensuring that the drawings, description and claims meet all formal requirements.

(iv) Utility or industrial applicability.

(v) Novelty [newness].

(vi) Non-obviousness or inventive step.

Four criteria, *viz.*, patentable subject matter, novelty, non-obviousness and sufficient disclosure are sufficient for granting patent.

16. When does a patent become operational?

The patent becomes effective immediately after its registration.

17. What is a patent infringement?

Infringement of the patent refers to an unauthorized use of the process or product by other person. A patent owner may recover money damages in the form of a reasonable royalty. If the infringement is found willful, three times money is recovered.

18. What are the Rights Protected by a utility patent?

A utility patent once granted, provide the owner of the patent the right to prevent others from making, using, selling and importing a product covered by the utility patent.

19. What are the Requirements for filing a patents?

Four criteria, *viz.*, patentable subject matter, novelty, usefulness and non-obviousness are sufficient for granting patent.

20. What are the differences between a patent and trademark?

There are several differences between patent and trademarks which have been presented in Table 3.1.

TABLE 3.1: Comparison of Patents and Trademarks

Sl.No.	*Particulars*	*Patents*	*Trademarks*
1	Requirement	Novelty and Utility	New & distinctive
2	Period	15-20 years	Renewable indefinitely
3	Registration	Compulsory	Is compulsory
4	Validity	In original country	All Countries
5	Fee	Fees is charged	Fee is charged
6	Legal rights	There are legal Rights	Legal rights are there
7	Transfer	Admissible	Possible
8	Items covered	Products/process/invention/ innovation	Names, images & short phrases
9	Enforcement	After Registration	After registration
10	Control	By inventor	By owner
11	Infringement	Possible	Possible

21. What are the advantages of patents?

There are three main advantages of patents which are given below:

(i) It excludes competitors.
(ii) It builds licensing portfolio.
(iii) It enhances product value.

22. What are the disadvantages of patents?

There are four main disadvantages of patents which are given below:

(i) It involves cost of prosecution and renewal.
(ii) It requires public disclosure of the invention.
(iii) The validity is for a short duration [15-20 years].
(iv) Patents are territorial in nature.

Chapter 4

Trademark Rights

1. What is a trademark?

A trademark is a distinctive sign of some kind which is used by an individual, business organization or other legal entity to uniquely identify the source of its products and/or services to consumers, and to distinguish its products or services from those of other entities. Important points about trademarks are listed below:

(i) A trademark is a type of intellectual property. In other words, it a type of industrial property.

(ii) A trademark can be a word, name, symbol, device or mark which is used to identify and distinguish the goods or services of one company from goods or services of another.

(iii) Trademark is used to identify its product and to distinguish them from others. It is the name of a product made by a particular person or company.

(iv) The period of protection for a trademark varies, but can generally be renewed indefinitely.

(v) Á trademark indicates source of the product or acts as a badge of origin.

2. What is the duration of the protection of a trademark?

Initial term for a trademark registration is 10 years. It can last as long as the owner wants. Between the fifth and sixth year after the date of initial registration, the trademark registrant will need to file documents to keep the mark alive. A renewal then be filed every 10 years. Thus it can last indefinitely.

3. Is Registration compulsory for a trademark?

The registration is not essential for a trademark. However, it is advisable to register the trademark to get legal rights against infringement of trademarks. Almost all countries in the world register and protect trademarks. The prescribed Fee has to be paid for registration of trademark.

4. What is Validity of a trademark?

The trademark rights are National, not International. Therefore, registration in one country does no confer rights in other countries. A separate application will be needed for registration in other country.

5. Is it possible to Transfer a trademark?

The owner of the trademark has right to authorize another person to use it on payment (royalty) basis.

6. What are the Matters Covered by a trademark?

A trademark includes a distinctive word, phrase, logo, domain name, graphic symbol, slogan, image or other device used to identify the source of a product from others. Thus a variety of items are covered by trademark. Recently colors [such as John deere green], sounds and scents have also been registered as trademarks.

7. Who has control over the trademark?

A trademark owner has rights to prevent unauthorized use of trademark.

8. When does a trademark become effective?

The trademark becomes effective when it is first publicly put to continuous commercial use, even without registration.

9. What is Infringement of a trademark?

An unauthorized use of a trademark by a person other than the trademark owner is called infringement of trademark. Infringement occurs when someone uses the same or similar trademark for a similar good or service. Infringement does not require the infringer to be competitor, but infringement does require a similar product.

10. What are the Rights Protected by a trademark?

Registration of trademarks provides exclusive right to the owner of the trademark to use it to identify goods or services.

11. What are the Requirements for registering a trademark?

Trademarks apply to products and services. There are some requirements for registration of a trademark. It must be distinctive, not descriptive and there must be no earlier rights or multiple appearance. Moreover, it should not confuse with other trademark. It should not be prohibitive (reserved) and generic.

12. What do you mean by service mark ?

There are many types of trademarks such as, (1) service mark, (2) trade dress, (3) collective mark and (4) certification mark. A trademark that is used in relation to services rather than products is called a service mark. It promotes services or events. Some examples of service marks are as follows:

(i) Amazon.com [retail website].

(ii) B.Jack in the Box [fast food service].

(iii) Kinko's [photocopying service].

(iv) Blockbuster [video rental service].

(v) Fed Ex logo delivery service].

13. What a trade dress?

A trademark that is known by its distinctive packaging is referred to as trade dress, for example the yellow packaging of Kodak film. The trade dress often serves the same function as a trademark or service mark.

14. What is a collective mark?

A trademark used by members of a group or organization is known as collective mark. It may be a symbol, label, phrase or other mark. These are used to identify goods, products or services rendered by a group. Collective marks are often used to show membership in a union, association or other organization.

15. What ia certification mark?

A certification mark is a symbol, name or device used by an organization to guarantee for the quality products and services provided by others such as Good Housekeeping Seal of Approval.

Based on the relationship, between the mark and the underlying product, trademarks are of four types,*viz.* (1) arbitrary, (2) suggestive, (3) descriptive, and (4) generic. These are defined as under.

16. Define an arbitrary mark.

An arbitrary mark is one that bears no logical relationship to the underlying product. For example, the words Ëxxon"and "Kodak"bear no inherent connection to their underlying products. It is also known as fanciful mark.

17. What is suggestive mark?

This is a mark that evokes or suggests a characteristic of the underlying good. For example the word "Copper tone" suggests the color of a deep sun tan.

18. What is meant by descriptive mark?

A descriptive mark directly describes rather than suggests, a characteristic or quality of underlying product such as its color, odor, function, diamensions or ingradients. For example "Holiday Inn"and A"ll Bran"describe hotel rooms and breakfast cereal.

19. What is a generic mark?

A generic mark is a mark that describes the general category to which the underlying product belongs. For example, the term" Computer" is a generic term for computer equipment. Such terms are not entitled for protection.

20 Describe the Procedure of Filing Trademarks.

The application for registration of trademark is to be submitted in the Patent and Trademark Office. The prior use of the trademark for commercial use is required for registration. The registration process consists of the following steps:

(1) **Filling of Application:** The prescribed application form is used and the application has to be carefully completed and signed. The following documents should be attached with the application.

  (i) Drawing of the mark on separate piece of paper.

  (ii) Prescribed Fee: The Fee is not refundable even if the application is rejected.

  (iii) **Specimens**: The manner in which the mark is to be used should be indicated. The application must clear reproduction of the sign filed for registration including, any colors, forms or three. dimensional features. The application must also contain a list of goods or services to which the sign would apply. The sign must be distinctive to identify the product from other products. Each application must include three specimens.

(2) **Examination of the application:** The application is examined in the PTO by an examining attorney. If the application is found correct, the attorney will approve the mark for publication in the official Gazette. A trademark search may also be conducted to know whether the mark is already in the use by some other party.

(3) **Opposition:** If any party thinks that the mark will damage it rights, opposition can be filed within 30 days from the date of publication in the Gazette.

(4) **Registration:** If no opposition is filed, the PTO will register the trademark and issue a registration certification in a reasonable time.

21. What are the differences between Trademarks and Trade Secret ?

There are several differences between trademarks and trade secret which have been presented in Table 4.1.

TABLE 4.1: Comparison of Trademarks and Trade Secret

Sl.No.	*Particulars*	*Trademarks*	*Trade Secret*
1	Requirement	New and distinctive	Any thing confidential of commercial use
2	Period	Renewable indefinitely	Indefinite duration
3	Registration	Is compulsory	Not required
4	Validity	In original country	With the company concerned
5	Fee	Fee is charged	No Fee

Sl.No.	*Particulars*	*Trademarks*	*Trade Secret*
6	Legal rights	Legal rights are there	No legal rights
7	Transfer	Possible	Possible
8	Items covered	Names, images and short phrases	Formula, pattern, device , business methods, source of supply, customers list, *etc.*
9	Enforcement	With commercial use	With commercial use
10	Control	By owner	By owner
11	Infringement	Possible	Not possible

22. What are the Advantages of Trademarks ?

There are several advantages of trademark which are presented below:

(i) Excludes competitors

(ii) Renewable indefinitely

(iii) Can be protected by common law rights

(iv) Enhances reputation of the company

23. What are the Disadvantages of Trademarks ?

There are three main disadvantages of trademark which are presented below:

(i) It involves cost of prosecution and renewal.

(ii) Infringement is possible.

(iii) Registered in the country where product is used.

Chapter 5

Rights Related to Industrial Design

1. What is a design?

A design is defined as the features of shape, configuration, pattern or ornament [or any combination thereof] applied to a finished article by any industrial process or means[whether manual, mechanical or chemical or combined]. They are solely judged by the eye. For example, it may the shape of a table, or the ornamentation on the handle of a spoon. The article may be made by hand, tool or machine.

2. What is an industrial design?

An industrial design refers to any original shape, picture, or some combination applied to a useful article of manufacture. In other words, it refers to the design of the mass-produced products of our every day environment, from sinks and furniture to computers. Industrial designs optimize the function, value, and appearance of products and systems for the mutual benefit of both user and manufacturer.

3. What is Indian Law of design?

The Indian law of design is preserved [enshrined] in the Designs Act 2000. The Act provides for the registration of designs in India. The rights granted under the Act are operative in the whole of India. It protects only those designs that are aesthetic in nature. They may be decorative elements added to the article or they may be part of very shape or configuration. Novelty and originality are important criteria in a design for registration. In addition, only those designs that are applied to an article by an industrial process are protected.

4. What is the duration of protection of a design?

Initially the design is registered for a term of 5 yearly basis. Later on it can be renewed on 5 years by paying the prescribed Fee. In some cases, it may be registered up to 15 years.

5. Is Registration compulsory for a design?

Registration is essential to get legal rights. There is no protection without registration.A nominal processing Fee is charged for registration.

6. What is the Validity of a design?

The protection of the design is valid only in the country where it has been registered. The protection in other countries can be obtained by filling separate application in each country.

7. Is authorization of a design possible?

A registered design is a property like any other commodity which can be purchased, sold, hired or licensed. It is advisable to register the sale of rights in the registration office before use of the design by new owner.

8. What are the Matters Covered by Industrial designs?

Industrial designs include any original shape, picture, or some combination applied to a useful article of manufacture. Industrial designs do not cover following things.

a. The functional features of an article.

b. A principle of construction, or how an article is built.

c. The materials used in the construction of an article.

d. Colour per se.

e. Ideas.

9. Who has the control over industrial designs?

The owner has the legal rights to prevent others from unauthorized use of his design.

10. When does an industrial design come into force?

The industrial design comes into force immediately after it registration in the patents and trademarks office.

11. What is the infringement of industrial design?

Unauthorized use of a design is called infringement of such design. The owner has the right to take legal action against the infringer and claim damages. However, the legal action must be taken within three years of the alleged offence.

12. What are the rights protected by industrial designs?

It provides exclusive rights to the owner such as to make,imports, sell or hire out or allow others to use the design under terms settled with the registered owner.

13. What are the requirements for registration of an industrial design ?

There are two requirements of industrial design for registration, *viz.* novelty and commercial application. In other words, it should be different from previously registered designs and it should not be made public before registration. It is made open to public after registration.

14. What is the procedure for registration of an industrial design ?

The registration is done in Patents and Trademarks Office. The owner of the design has to complete the following formalities:

1. He has to submit duly filled up application in prescribed form in the Patents Office.
2. Prescribed Fee has to paid along with application.
3. The specimen of the design is to be provided. The only visual aspects of the design such as shape, configuration, pattern or ornament or any combination thereof should be described. It must state clearly what the design is. Too much details are not required.
4. The registry office will make a search about novelty of the design.
5. If requirements are met, a certificate of registration is issued otherwise registration will be refused which is appealable.
6. Initially 5 years registration is granted that can be renewed on five yearly basis.

Normally, the registration process should be completed within six months. However, a total of 12 months [extendable to 15 months] time is allowed for completion of registration process. The design should not be made public before registration otherwise registration may later be proved invalid.

15. What are the similarities and differences between an industrial design and patent?

There are some similarities and differences between industrial design and patent which have been presented in Table 5.1 below:

TABLE 5.1: Comparison of Industrial Design and Patents

Sl.No.	*Particulars*	*Industrial Design*	*Patents*
1	Requirement	A design should be novel	It should be novel
2	Period	5-15 years	15-20 years
3	Registration	Required	Compulsory
4	Validity	In the country of registration	In original country
5	Fee	Fee is charged	Fees is charged
6	Legal rights	Owner has legal rights	There are legal Rights
7	Transfer	Possible	Admissible
8	Items covered	Feature of shape, pattern, configuration or ornament on a finished article.	Products/process/invention// innovation
9	Enforcement	After registration	After Registration
10	Control	By right owner	By inventor
11	Infringement	Possible	Possible

16. What are the differences and similarities between an industrial design and copyright?

The differences and similarities between industrial design and copyrights are presented in Table 5.2.

TABLE 5.2: Comparison of Industrial Design and Copyright

Sl.No.	*Particulars*	*Industrial Design*	*Copyrights*
1	Requirement	A design should be novel	Originality and wide utility
2	Period	5-15 years	70-120 years
3	Registration	Required	Not compulsory
4	Validity	In the country of registration	All countries
5	Fee	Fee is charged	No Fee
6	Legal rights	Owner has legal rights	No legal Rights
7	Transfer	Possible	Possible
8	Items covered	Feature of shape, pattern, configuration or ornament on a finished article.	Literary, music, movie, *etc.*
9	Enforcement	After registration	Immediately after release of work
10	Control	By right owner	By Author/creator
11	Infringement	Possible	Very common

17. What are the differences and similarities between an industrial design and trademarks?

The differences and similarities between industrial design and trademarks are given in Table 5.3.

TABLE 5.3: Comparison of Industrial Design and Trademarks

Sl.No.	*Particulars*	*Industrial Design*	*Trademarks*
1	Requirement	A design should be novel	New and distinctive
2	Period	5-15 years	Renewable indefinitely
3	Registration	Required	Is compulsory
4	Validity	In the country of registration	In original country
5	Fee	Fee is charged	Fee is charged
6	Legal rights	Owner has legal rights	Legal rights are there
7	Transfer	Possible	Possible
8	Items covered	Feature of shape, pattern, configuration or ornament on a finished article.	Names, images and short phrases
9	Enforcement	After registration	With its commercial use
10	Control	By right owner	By owner
11	Infringement	Possible	Possible

18. What are the differences and similarities between an industrial design and trade secret?

The differences and similarities between industrial design and trade secret are presented in Table 5.4.

TABLE 5.4: Comparison of Industrial Design and Trade Secret

Sl.No.	*Particulars*	*Industrial Design*	*Trade Secret*
1	Requirement	A design should be novel	Any confidential knowledge of commercial use
2	Period	5-15 years	Indefinite duration
3	Registration	Required	Not required
4	Validity	In the country of registration	With the company concerned
5	Fee	Fee is charged	No Fee
6	Legal rights	Owner has legal rights	No legal rights
7	Transfer	Possible	Possible
8.	Items covered	Feature of shape, pattern, configuration or ornament on a finished article.	Formula, pattern, device, business methods, source of supply, customers list, *etc.*
9	Enforcement	After registration	With commercial use
10	Control	By right owner	By owner
11	Infringement	Possible	Not possible

19. What are advantages of industrial design?

Industrial Designs:

1. Enhance aesthetic value of the product.
2. Provide legal rights to the owner against unauthorized use by others.
3. Make an article attractive and increase it marketability.
4. Can be purchased, sold or licensed.
5. Enhance reputation of the company.
6. Encourage creativity in the industrial sector and in arts and crafts.

20. What are disadvantages of industrial design?

Industrial Designs:

1. Require high technical skill.
2. Are liable to infringement.
3. Involve renewal cost.
4. Are valid in the country of registration only.
5. Lead to monopoly.

Chapter 6

Rights Related to Utility Models

1. What is a utility model?

A utility model is an exclusive right granted for an invention, which allows the right holder to prevent others from commercially using the protected invention, without his authorization, for a limited period of time. In its basic definition, which may vary from one country (where such protection is available) to another, a utility model is similar to a patent. In fact, utility models are also referred to as "petty patents" or "innovation patents, minor patent and small patent."

2. Describe important points related to utility models.

The important points about utility model are listed below:

(i) Utility model is an intellectual property rights to protect inventions.

(ii) Utility models are also known as petty patents, petty inventions, petty innovations, utility innovations, minor patents and small patent.

(iii) The rights conferred by utility models are similar to those granted by patent laws but has a shorter term.

(iv) They are registered with National Patent Office.

(v) They are more suited to incremental inventions.

(vi) Presently, utility models are protected in about 170 countries of the World.

(vii) In some countries, utility model protection can only be obtained for certain fields of technology and only for products but not for processes.

(viii) Utility models are considered particularly suited for "minor" improvements to, and adaptations of, existing products. Utility models are primarily used for mechanical innovations.

(ix) Utility model system has been established for protection of small devices.

(x) Utility models Act provides a "low-cost entry point into the intellectual property system.

(xi) Utility Model Act defines a device as "a creation of technical ideas by which a law of nature is utilized." The words "inventions" and "devices" are used for ideas involving different levels of creativity.

(xii) Some countries exclude particular subject matter from utility model protection. For example, methods, plants and animals are normally barred from utility model protection

3. What are the Materials Covered by utility models?

Utility model", means any form, configuration or disposition of elements of some appliance, working tools and implements as articles of everyday use, electrical and electronic circuitry of other object or part thereof in so far as they are capable of contributing some benefit or new effect of saving in time, energy, or labour or improving the hygienic or socio-physiological working conditions by means of new configuration, arrangement or device or a combination thereof and are industrially applicable.

4. List some Countries Where utility models are Protected.

A utility model is an intellectual property right to protect inventions. This right is available in a number of national legislations, such as Australia, Argentina, Armenia, Austria, Belarus, Belgium, Brazil, Bulgaria, China, Colombia, Costa Rica, Czech Republic, Denmark, Estonia, Ethiopia, Finland, France, Georgia, Germany, Greece, Guatemala, Hungary, Ireland, Italy, Japan, Kazakhstan, Kenya, Kyrgyzstan, Malaysia, Mexico, Netherlands, OAPI, Peru, Philippines, Poland, Portugal, Republic of Korea, Republic of Moldova, Russian Federation, Slovakia, Spain, Tajikistan, Trinidad and Tobago, Turkey, Ukraine, Uruguay and Uzbekistan. It is very similar to the patent, but usually has a shorter term (often 6 or 10 years) and less stringent patentability requirements.

5. What are the Requirements for Registration of a utility model?

Most countries having utility model laws require that the invention be new. However, many patent or utility model offices do not conduct substantive examination and merely grant the utility model after checking that utility model applications comply with formalities. The protection for utility models shall be available for utility models which are new and have industrial applications. Each utility model application shall be subjected to substantive examination for compliance with substantive examination requirements. Each designated state prior to registration shall be accorded the opportunity to give an opinion relating to registration in its territory of the utility model.

6. Describe the procedure for registration of utility model.

The registration of a utility model consists of three main stepts, *viz.*, (i) filing application, (ii) examination of the application, and publication and registration.

(i) **Filing Application:** Utility model applications may be prepared and filed at local patent offices in countries where utility model protection is available. The application for a utility model should contain (i) name of the applicant, (ii)description of the utility model, (iii) drawings of the model, (iv) abstract, (v) state for which utility model is to be protected, and (vi) prescribed fee.

Conversion of Application: Application refused by the office or application which have been refused consideration can on request be converted into national applications. Patent application can also be converted into a utility model application or vice versa on request and such a converted application can benefit from the filing date of the initial application.

(ii) Examination of the Application: The application is examined by the experts. There is a pre-registration examination for Utility Models. This is conducted in three stages. The first stage involves examination to determine whether the application meets the requirements.

The Second Stage involves substantive examination. The substantive examination is carried out to determine whether the utility model is industrially applicable and secondly whether it is new. Novelty in this case shall be the local Novelty in a designated state. On completion of the second stage, the application is forwarded to designated states. The Third Stage involves examination by a designated state for novelty and for conformity with the national laws. This examination or verification should be completed within six months from the date of notification

Each designated state is required to carry an examination to ascertain whether the Utility Model is New in the territory of the state designated.

(iii) Registration and Publication: Registration and Publication entails recording of the Utility Model in the Journal and the Register. It also entails issuance of a Certificate and Publication of a copy of the specification.

(iv) Registration Fee: The registration fee and maintenance charges are much lower than those for patents.

7. What are the rights covered by utility model?

Under the Utility Model Act, "the owner of a utility model shall have an exclusive right to commercially work the registered model" and is allowed to demand an injunction, compensation for damage, and compensation for unfair profits gained by the infringer.

8. Is authorization possible for utility models?

The inventor has the right to authorize another person to make use of his invention on royalty basis.

9. What the duration for protection of utility models?

The utility model is registered for a shorter period than patent. The period of protection generally varies from 6 to 10 years in different countries.

10. What is the Validity of utility models?

The utility model will be valid in the country where it has been registered.

11. What is Infringement of utility models?

Infringement of utility model is possible. The owner or the inventor has legal rights to sue the infringer and recover damaged caused by way of infringement.

12. What are the differences between patent and utility model?

There are several differences between patents and utility models. The main differences between them are as follows:

(i) The requirements for acquiring a utility model are less strict than for patents. Novelty and non-obviousness" may be much lower or absent altogether. In practice, protection for utility models is often sought for innovations of a rather incremental character which may not meet the patentability criteria.

(ii) The term of protection for utility models is shorter than for patents and varies from country to country (usually between 6 and 10 years without the possibility of extension or renewal).

(iii) In most countries where utility model protection is available, patent offices do not examine applications as to substance prior to registration. This means that the registration process is often significantly simpler and faster, taking, on average, six months.

TABLE 6.1: Differences between Utility Models and Patent

Sl.No.	*Particulars*	*Utility Model*	*Patents*
1	Period	6-10 years	15-20 years
2	Registration	Is compulsory but easy	Compulsory and difficult
3	Fee	Less Fee is charged	More Fees is charged
4	Items covered	Utility inventions	Products/process/invention/ innovation
5	Protection of plants and animals	Not permitted	Permitted

(iv) Utility models are much cheaper to obtain and to maintain.

(v) The Utility Model Act extends protection only to "devices relating to the shape or construction of articles or combination of articles." Any methods such as a manufacturing process are not covered by the Utility Model Act.

(vi) In the patent system, a substantive examination principle has been introduced. In the utility model system, on the other hand, a non-substantive examination principle has been introduced.

(vii) In addition to above, patent and utility models differs differ from each other in the durations of rights and administration fees.

13. What are the similarities between patent and utility model?

There several similarities between patents and utility models which are presentedin Table 6.2.

TABLE 6.2: Similarities between Utility Models and Patent

Sl.No.	*Particulars*	*Utility Models*	*Patent*
1	Requirement	Novelty	Novelty
2	Validity	In country of registration	In original country
3	Legal rights	Legal rights are there	Legal Rights are there
4	Authorization	Admissible	Admissible
5	Enforcement	After registration	After Registration
6	Control	By Inventor	By inventor
7	Infringement	Possible	Possible

14. What are the Advantages of Utility Models?

There are several advantages of utility models which are listed below:

(i) It includes simple devices.

(ii) Registration procedure is simple.

(iii) Registration is faster.

(iv) Much cheaper to obtain and maintain.

15. What are the Disadvantages of Utility Models?

There are some disadvantages of utility models which are listed below:

(i) It protects for shorter duration [6-10 years].

(ii) Generally it cannot be renewed.

(iii) Plants and animals cannot be registered.

(iv) Processes are not protected.

Chapter 7

Geographical Indication

1. What is geographical indication?

Geographical Indication may be defined as the name of a region, a specific place or, in exceptional cases a country, used to describe an agricultural, natural or manufactured goods, product or a food stuff. In other words, a geographical indication is a sign used on goods that have a specific geographical origin and often possess qualities or a reputation that are due to that place of origin. In other words, a geographical indication is a name or sign used on certain products or which corresponds to a specific geographical location or origin (*e.g.* a town, region, or country).

2. What is Indian Legislation for geographical indication?

The World Intellectual Property Organization (WIPO) helps in protection of Geographical Indications [GI]. About 170 geographical indications have been registered by Lisbon Agreement members up to 1997. Now this number would be much more. In India, the Parliament has passed the Geographical Indications of Goods (Registration and Protection) Act, 1999 [in December 1999]. This Act seeks to provide for the registration and better protection of geographical indications relating to goods in India. The Act would be administered by the Controller General of Patents, Design and Trade Marks who is the Registrar of Geographical Indications. The Geographical Indications Registry is located at Chennai.

3. How many geographical indications have been registered so far?

Till 1997, 170 geographical indications have been registered in some advanced countries because they have developed the system of protecting GI. In the past, wines, spirits, cheeses, tobacco, which account for 88 per cent, have been registered as GI. Wines and spirits account for almost 71 per cent of all registrations. In addition to wines and spirits, there is need for protection of agricultural products as GI. For this there is need to establish a multilateral system of identification and registration of GI. Geographical Indication may be an agricultural, natural or manufactured goods or product.

4. What is the duration for protection of geographical indication?

The registration of a geographical indication is valid for a period of 10 years. It can be renewed from time to time for further period of 10 years each. If a registered geographical indication is not renewed it is liable to be removed from the register.

5. Is registration compulsory for protection of geographical indication?

The registration is not compulsory. The registration provides better legal protection to take action against infringement. In other words, it confers legal protection to geographical indications, and prevents unauthorized use of a registered geographical indication by others.

6. What is the validity of geographical indication?

The registration of the geographical indication is valid in the country of its origin.

7. Is it possible to transfer geographical indication?

A geographical indication cannot be sold, purchased, transferred, licensed, pledged, mortgaged or put to such other agreement.

8. What are the matters covered geographical indication?

It is used to identify agricultural, natural and manufactured goods.

9. Who has the control over geographical indication?

An authorized user has the exclusive rights o use the geographical indication in relation to goods for which it is registered.

10. When does geographical indication become effective?

The protection of geographical indication becomes effective after its registration.

11. What is the Infringement of geographical indication?

Infringement of a geographical indication is possible. The registered proprietor and authorized users can initiate action against infringement.

12. What are the rights protected by geographical indication?

The proprietor or authorized user has the rights to use the geographical indication and prevent other from using such material without permission.

13. What are the requirements for registration of a geographical indication?

It originates from a definite geographical territory and should have a special quality or reputation or other characteristics. The main features are as given below:

(i) It is an indication.

(ii) It originates from a definite geographical territory.

(iii) It is used to identify an agricultural, natural or manufactured goods/ product.

(iv) The manufactured goods should be processed or prcessed in that territory.

(v) It should have a special quality or reputation or other characteristics.

14. What is the procedure for registration of geographical indication?

The registration is done by the registrar of Geographical Indications. For registration any association of persons, producers, organization or authority established by or under the law can apply The applicant must represent the interest of the producers. The registration consists of following steps:

(i) **Filing Application:** The application in the prescribed form is filed to the registrar of Geographical Indications along with prescribed Fee and other details.

(ii) **Examination:** The application is examined by the registrar office by conducting search. The application is accepted, if it fulfils the requirements. Then it is advertised in the GI journal for objection if any.

(iii) **Acceptance:** If there is no objection, the application is accepted for grant of registration.

(iv) **Registration:** Then the GI is entered in the Register. The register of geographical indication is divided into two parts, *viz.* part A and part B. Part A consists of particulars relating to registered geographical indication and part B consists of particulars of the registered authorized users.

(v) **Issue of Certificate:** Finally a certificate is issued to the registered user. The registration of GI is for a period of 10 years which is renewable for further periods of 10 years each.

15. What are the differences between geographical indication and patents?

There are several differences between geographical indication and patents which have been presented in Table 7.1.

TABLE 7.1: Differences between Geographical Indication and Patents

Sl.No.	*Particulars*	*Geographical Indication*	*Patents*
1	Requirement	Reputed product of a specific place or territory	Novelty and Utility
2	Period	10 years but renewable	15-20 years
3	Registration	Not compulsory	Compulsory
4	Legal rights	After registration only	There are legal Rights
5	Transfer	Not allowed	Admissible
6	Items covered	Agricultural, natural and manufactured goods	Products/process/invention// innovation
7	Control	By authorized user or proprietor	By inventor

16. What are the similarities between geographical indication and patents?

There are some similarities between geographical indication and patents which have been presented in Table 7.2.

TABLE 7.2: Similarities between Geographical Indication and Patents

Sl.No.	*Particulars*	*Geographical indication*	*Patents*
1	Validity	In the country of registration	In original country
2	Fee	Fee is charged if registered	Fees is charged
3	Enforcement	After registration	After Registration
4	Infringement	Possible	Possible

17. What are the differences between geographical indication and trademarks?

There are several differences between geographical indication and trademarks as given below:

(i) **Number:** The total number of registered trademarks is more than 16 millions. However, only about 6 millions trademarks are currently in use worldwide. Whereas about 170 GI have been registered so far. Thus only limited number of trademarks are useful

(ii) **Beneficiaries:** In case of GI, only local people or people of a region are beneficiaries, whereas in case of trademarks large number of people are beneficiaries.

(iii) **Role of State:** The State Government plays important role in protection of. Trademarks are promoted by individual company.

(iv) **Production:** The production of GI is collective by producers of a place or region, whereas the production of trademarks is by individual company.

(v) **Protection:** The protection of GI is difficult as compared to trademarks.

(vi) **Creation:** The creation of new trademarks is easier than creation of new GI.

Besides these, there are several other differences between geographical indications and trademarks which are presented in Table 7.3.

TABLE 7.3: Differences between Geographical Indication and Trademarks

Sl.No.	*Particulars*	*Geographical Indication*	*Trademarks*
1	Requirement	Reputed product of a specific place or territory	New and distinctive
2	Period	10 years but renewable	Renewable indefinitely
3	Registration	Not compulsory	Is compulsory
6	Legal rights	After registration only	Legal rights are there
7	Transfer	Not allowed	Possible
8.	Items covered	Agricultural, natural and manufactured goods	Names, images and short phrases

Sl.No.	*Particulars*	*Geographical Indication*	*Trademarks*
9	Enforcement	After registration	With its commercial use
10	Control	By authorized user or proprietor	By owner

18. What are the similarities between geographical indication and trademarks?

There are some similarities between geographical indication and trademarks as given in Table 7.4.

TABLE 7.4 Similarities between Geographical Indication and Trademarks

Sl.No.	*Particulars*	*Geographical indication*	*Trademarks*
1	Validity	In the country of registration	In original country
2	Fee	Fee is charged if registered	Fee is charged
3	Infringement	Possible	Possible

19. What are advantages of geographical indication?

There are several advantages of geographical indications which are presented below:

(i) GI protection provides legal rights to the users.
(ii) It prevents unauthorized use of registered GI.
(iii) It boosts exports.
(iv) It promotes economic prosperity of producers of goods in a geographical territory.
(v) Geographical indications bring name to place of their origin or territory.

20. What are disadvantages of geographical indications?

There are some disadvantages of geographical indications which are listed below:

(i) It prevents others for use.
(ii) It leads to monopoly.
(iii) Users are limited to place of origin.

21. What are examples of Indian geographical indications?

In India some geographical indications have been identified which are listed below:

(i) Basmati Rice.
(ii) Darjeeling Tea.
(iii) Kanchipuram Silk Saree.
(iv) Alphanso Mango.

(v) Nagpur Orange.

(vi) Kolhapuri Chappal.

(vii) Bikeneri Bhujia.

(viii) Agra Petha.

(ix) Banarasi Saree

The above geographical indications can be classified into three groups, *viz.* (i) agricultural, (ii) natural, and (iii) manufactured

(i) Agricultural GI: This group includes Basmati Rice, Nagpur Orange and Alphanso mango.

(ii) Natural GI: This group includes Darjeeling Tea.

(iii) Manufactured GI: This group includes Kanchipuram Silk Saree, Banarasi Silk Saree, Kolhapuri Chappal, Bikaneri Bhujia and Agra Petha.

Chapter 8

Trade Secret

1. What is a trade secret?

Trade secret may be a formula or process or device used in business that is not published or divulged which gives an advantage over competitors. It is, known only to its owner, that is used in producing something of commercial value.

2. What are main points related to trade secret?

The main points related to trade secret are briefly presented as follows:

(i) Trade secret is a knowledge that is possessed by an individual or institution and kept secret to prevent others from using the knowledge in a competitive manner.

(ii) There is no specific period for trade secret. It may continue life long or for generations together.

(iii) There is no need of registration for trade secret. Trade secret can never be patented, because it does not publicize the knowledge.

(iv) It does not provide opportunity to others for improvement of innovation.

(v) It is not applicable to books, equipments, plant varieties, designs which are openly used.

3. What is the duration for protection of trade secret?

It can be protected for indefinite period. In other words, it can be protected so long as the idea/formula is kept secret.

4. Is registration compulsory for a trade secret?

There is no need of registration for a trade secret.

5. What is the Validity of a trade secret?

It is valid with the company concerned.

6. Is it possible to transfer a trade secret?

The owner of the trade secret can reveal the secrecy to other person on payment basis. Trade secret is a property right. Hence, the owner of the trade secret can sell his trade secret to someone else. Moreover, the owner can license his trade secret to others on royalty basis.

7. What are the matters covered by trade secret?

A trade secret may consist of following things:

(i) A formula for a chemical compound
(ii) A process of manufacturing, trating opr preserving materials
(iii) A pattern for a machine or other device or
(iv) A list of customers
(v) Novelty and inventions are not requisite for a trade secret as they are not for patentability.

8. Who has the control over the trade secret?

The trade secret is controlled by the owner of the company. Trade secrets protect the application of ideas.

9. When does a trade secret become effective?

The trade secret becomes effective when the product is put to continuous commercial use without disclosure.

10. Infringement of a trade secret is possible?

The infringement of the trade secret is possible through reverse engineering.

11. What are the rights protected by a trade secret?

Protection of trade secret provide the benefit of production and marketing only to the owner of the trade secret. The owner of a trade secret has the right to prevent use or dissemination of the trade secret by any one who learned or derived the trade secret from the owner. The owner has no trade secret rights against any one who independently discovers the trade secret.

12. What are the requirements for a trade secret?

Any thing confidential and of commercial use can be protected by trade secret.

13. How a trade secret can be protected?

A trade secret can be safeguarded in various ways such (i) by obtaining written agreement from the employees, who know about the trade secret, not to disclose the secret to anyone and (ii) by keeping the documents, containing information about trade secret, in the locking system. The employer may recover damages against such a person [employee] revealing the secret. The written agreements are of two types, *viz.* (a) non-disclosure agreements, and (b) non-compete agreements.

14. What are the differences between patents and trade secret?

There are several differences between patent and trade secret which have been presented in Table 8.1.

TABLE 8.1: Differences between Patent and Trade Secret

Sl.No.	*Particulars*	*Patents*	*Trade Secret*
1	Requirement	Novelty	Any thing confidential of commercial use
2	Period	15-20 years	Indefinite duration
3	Registration	Compulsory	Not required
4	Validity	In original country	With the company concerned
5	Fee	Fees is charged	No Fee
6	Legal rights	There are legal Rights	No legal rights
7	Transfer	Admissible	Possible
8.	Items covered	Products/process/invention// innovation	Formula, pattern, device , business methods, source of supply, customers list, *etc.*
9	Enforcement	After Registration	With commercial use
10	Control	By inventor	By owner
11	Infringement	Possible	Not possible

15. What are the differences between trade secret and copyrights?

There are several differences between trade secret and copyrights which have been presented in Table 8.2.

TABLE 8.2: Differences between Trade Secret and Copyrights

Sl.No.	*Particulars*	*Trade Secret*	*Copyrights*
1	Requirement	Any thing confidential and of commercial use	Originality and wide utility
2	Period	Indefinite duration	60-110 years
3	Validity	With the company concerned	All countries
4	Items covered	Formula, pattern, device , business methods, source of supply, customers list, *etc.*	Literary, music, movie, *etc.*
5	Enforcement	With commercial use	Immediately after release of work
6	Infringement	Not possible	Very common

16. What are the similarities between trade secret and copyright?

There are some similarities between trade secret and copyrights which have been presented in Table 8.3.

TABLE 8.3: Similarities between Trade Secret and Copyright

Sl.No.	*Particulars*	*Trade Secret*	*Copyrights*
1	Registration	Not required	Not compulsory
2	Fee	No Fee	No Fee
3	Legal rights	No legal rights	No legal Rights
4	Transfer	Possible	Possible
5	Control	By owner	By Author/creator

17. What are advantages of trade secret ?

There are several advantages of trade secret which are listed below:

(i) It is kept in strict confidence.

(ii) In can be protected for indefinite duration.

(iii) It provides competitive business advantage.

(iv) It does not require registration.

18. What are disadvantages of trade secret?

There are some disadvantages of trade secret which are listed below:

(i) It is lost if secret is revealed.

(ii) The secrecy may be revealed by the employee.

(iii) use of trade secret for long period eliminates the possibility of patenting an invention.

(iv) It prevents further improvement in a secret process by others.

(v) The owner has monopoly in secret information.

Chapter 9

Related Rights

1. What are related rights?

Related rights is a term in copyright law, used in opposition to the term "authors' rights". The term neighbouring rights is exactly equivalent, and a more literal translation of the original French *droits voisins*. Related rights in civil law are rights which are similar to authors' right but which are not connected with the actual author of the work. Both authors' rights and related rights are copyrights in the sense of English or U.S. law. Main points about related rights are given below:

1. Related rights are similar to copyrights.
2. Related rights come under primary property rights.
3. Related rights are also known as neighouring rights.
4. Related rights are closer to copyrights but are not covered by the Berne Convention.
5. Related Rights are covered by Rome Convention, 1961.

2. What are the matters covered by related rights?

Related rights or neighboring rights cover performers rights, the rights of phonogram producers, publishers' rights in typographical arrangements, the rights of broadcasting organizations in relation to their broadcasts, film producers, database creators, photographers and designers. These rights are internationally protected by the Rome Convention for the Protection of Performers, Producers of Phonograms and Broadcasting Organisations signed in 1961. Within the European Union, the rights of film producers (as opposed to directors) and database creators are also protected by related rights, and the term is sometimes extended to include the *sui-generis* rights in semiconductor topologies and other industrial design rights.

3. What is the duration of Protection of related rights?

The period of protection varies with the matters protected by related rights and also with the countries. The period of protection as per Rome Convention for different matters is as follows.

TABLE 9.1: Period of Protection for different Matters

Sl.No.	*Items/Matters*	*Period (Years)*	*Date of Commencement*
1	Performers	15	From end of the year in which performance was made.
2	Phonogram producers	20	From end of the year in which phonogram was first published. It is 50 years in European Union.
3	Broadcasting organizations	20	From the end of the year in which the broadcast was first made.
4	Film Producers	50	From the end of the year in which the film was first published or otherwise made available to the public, or for fifty years from the end of the year in which the master copy was made if the film is not released.
5	Database creators	15	From the "completion" of the database.
6	Photographers	25	From date of creation.
7	Designers	25	After creation.
8.	Semiconductor designers	10-15	Ten years from the first commercial exploitation, or for fifteen years from the first creation.

4. What is the Relationship between related rights and authors' rights?

Related rights are independent of any authors' rights which might also exist in the work. Hence a CD recording of a song will be concurrently protected by four different copyright-type rights:

(i) The authors' rights of the composer of the music;

(ii) The authors' rights of the lyricist;

(iii) The performers' rights of the singer and the musicians; and

(iv) The producers' rights of the person or corporation which made the recording.

5. Whar are the rights covered by performers?

The performers have the right to prevent:

(i) The broadcast or communication to the public of their performance, unless this is made from a legally published recording of the performance;

(ii) The fixation (recording) of their performance; and

(iii) The reproduction of a recording of their performance.

6. What are the rights of phonogramme producers ?

The producers of phonogrammes, that is the person who makes the recording rather than the person who performs, has the right to prevent the direct or indirect reproduction of the recording.

7. What are the rights of broadcasting organizations?

Rome Convention specified that broadcasting organisations shall have the right to prohibit (or licence):

(i) The rebroadcasting of their broadcasts;
(ii) The fixation (recording) of their broadcasts;
(iii) The reproduction of fixations of their broadcasts; and
(iv) The communication of their broadcasts to the public in places where an entrance fee is charged.

8. What are the rights conferred with film producers?

Film producers have the right to prevent:

(i) The direct or indirect reproduction of the film (either the master copy or copies thereof);
(ii) The distribution of the film (or copies thereof) to the public for sale; and
(iii) The "making available to the public" of the film.

9. What are the rights of database creators?

Database creators have the right "to prevent extraction and/or re-utilization of the whole or of a substantial part, evaluated qualitatively and/or quantitatively, of the contents of that database."

Similarly photographers and designers have rights to prevent others to use their creations without permission.

10. Is authorization possible of related rights?

The owners of above mentioned property have the rights to authorise or to provide license to others to make use of their related rights on royalty basis. The performer has the right to licence:

(i) The distribution of recordings of their performance, for sale or other transfer of ownership;
(ii) The rental of recordings of their performances, unless there is a compulsory licence scheme in operation; and
(iii) The "making available to the public" of their performances, in effect their publication on the internet.

The owner of the phonogram has the rights to licence:

(i) The distribution of their phonograms, for sale or other transfer of ownership;
(ii) The rental of their phonograms, unless there is a compulsory licence scheme in operation; and
(iii) The "making available to the public" of their phonograms, in effect their publication on the internet.

Once a phonogram has been published, the producer cannot prevent its broadcast: an equitable fee for the licence may be either agreed between phonogram producers and broadcasters or imposed by law.

11. Is registration required for related rights?

Registration is not compulsory. However, registration provides legal rights to prevent others to make unauthorised use of the inventions or creations of the owners. The registration is done in the Copyrights Office and prescribed fee is charged.

12. When does the related rights become effective?

The related rights come into force after registration. Without registration, there is no protection of related rights.

13. How infringement of related rights can be prevented?

The unauthorized use of the protected matters is called infringement. The right holders can take legal action against those who are using their creations without permission. The penalties can be imposed against damages.

14. What are the international treaties for protection of related rights ?

Apart from the Rome convention, there are some other treaties which address the question of the protection of related rights. These treaties are as follows:

(i) Convention for the Protection of Producers of Phonograms Against Unauthorized Duplication of Their Phonograms (Geneva Phonograms Convention, 1971).

(ii) Convention Relating to the Distribution of Programme–Carrying Signals Transmitted by Satellite (Brussels Convention, 1974).

(iii) Treaty on Intellectual Property in Respect of Integrated Circuits (IPIC Treaty, 1989).

(iv) Agreement on Trade-Related Aspects of Intellectual Property Rights (TRIPS, 1994).

(v) WIPO Performances and Phonograms Treaty (WPPT, 1996).

These treaties do not grant related rights, but endorse that each country should provide related rights to the creators of original work. It will help in promoting artistic and creative work.

15. What are the differences between copyrights and related rights?

There are some differences between copyrights and related rights which are presented in Table 9.2.

TABLE 9.2: Differences between Copyrights and Related Rights

Sl.No.	*Particulars*	*Copyrights*	*Related Rights*
1	Period	70-120 years	10-50 years
2	Validity	All countries	In original country
3	Items covered	Literary, music, movie *etc.*	Performers, phonogram, database creators, film producers, designers, *etc.*

16. What are the similarities between copyrights and related rights?

Related rights are very similar to copyrights. The similarities between these two are presented in Table 9.3.

TABLE 9.3: Similarities between Copyrights and Related Rights

Sl.No.	*Particulars*	*Copyrights*	*Related Rights*
1	Requirement	Originality and wide utility	Originality
2	Registration	Not compulsory	Not compulsory
3	Fee	Fee is charged if registered	Fees is charged
4	Legal rights	Legal Rights only after registration	Legal Rights only after registration
5	Transfer	Admissible	Admissible
6	Enforcement	Immediately after release of work	After production or release or performance
7	Control	By Author/creator	By inventor
8	Infringement	Very common	Possible
9	Covered by	Berne Convention	Geneva Convention

17. What are advantages of related rights?

There are several advantages of related rights which are listed below:

(i) Registration is not compulsory.
(ii) It protects original work.
(iii) There is no fee for copyright protection.
(iv) It promotes creative work.
(v) It provides benefits to the creators.

18. What are disadvantages of related rights?

There are some disadvantages of related rights which are listed below:

(i) Protection is for shorter period than copyrights.
(ii) It does provide legal rights without registration.
(iii) Rights are valid only in the country where registered.
(iv) It may lead to monopoly.

Chapter 10

Infringement or Piracy

1. What is infringement?

Infringement refers to unauthorized copying or use of a protected form of intellectual property. An unauthorized use of material protected by copyright, patent or trademark law is called infringement. It is an invasion of one of the exclusive rights of intellectual property. It is the use or exploitation of another individual or organization's intellectual property rights without lawful authority. An unauthorized attempt to make, use, sell or have made a property right owned by another is called infringement.

2. What do you mean by piracy?

Piracy refers to an unauthorized duplication of goods protected by intellectual property law (*e.g.* copying software unlawfully). Copying or distributing protected works or products without the permission of the true owner/author of an intellectual property is called piracy. Piracy refers to illegal copying of a product such as software or music. The act of stealing or illegally distributing software is called piracy.

It is very clear that definitions of both infringement and piracy are similar and convey the same meaning. Hence, these are synonymous and interchangeable terms. However, infringements is in more usage in case of copyright, patents, trademarks and industrial design and piracy is commonly used in case of unauthorized use of software.

3. What is copyright infringement?

Infringement may be of different types such (i) copyright infringement, (ii) patent infringement, (iii) trademark infringe,(iv) industrial design, infringement, (v) biopiracy, and (vi) software piracy.

The illegal use of a copyrighted work or violation of the exclusive rights granted by the copyright law is referred to as infringement of copyrights. In other words, when another party besides the copyright owner reproduces a copyrighted work, in whole or in part, without the copyright owner's permission it is called copyright infringement. Infringement occurs when someone copies, distributes, publicly performs or displays copyrighted work without the permission of the artist or author of the work.

4. Define patent infringement.

Making, using or selling an invention protected by a patent without the inventor's permission is called patent infringement. In other words, infringement of a utility patent involves the making, using, selling, offering to sell, or importing of a patented product or process without permission.

5. What is trademark infringement ?

Violation of trade-mark rights through unauthorized use of a trade-mark is called trademark infringement. In other words, it refers to ihe improper use of another's trademark or a confusingly similar mark that creates the appearance of an affiliation with or actually being the other product or service.

6. Define industrial design infringement.

It refers to violation of industrial design rights through unauthorized use of a design or unauthorized use of industrial design without the permission of owner is called industrial design infringement.

7. What is Bio-piracy?

The illegal collection of indigenous plants by corporations who patent them for their own use is called bio-piracy. It refers to:

(i) Unauthorized use of biological resources such as plants, animals, micro-organisms and genes;

(ii) Unauthorized use of traditional communities' knowledge on biological resources,

(iii) Unequal share of benefits between a patent holder and the indigenous community whose resources and knowledge have been used; and

(iv) Patenting of biological resources with no respect to patentable criteria such as novelty, non-obviousness and usefulness.

Bio-piracy commonly operates through the application of Intellectual Property Rights to genetic resources and traditional knowledge. Bio-prospecting is the search for biological resources and related indigenous knowledge primarily for the purpose of commercial exploitation. The issue of bio-piracy is mostly raised by under-developed biodiversity rich countries such as India, Brazil and Malaysia among others.

8. What is software piracy?

Software piracy is the unauthorized duplication of computer software. In 2003, the estimated losses due to software piracy exceeded 29 billion dollars. Software piracy is of many forms such as (i) end user piracy, (ii) client-server overuse, (iii) internet piracy, (iv) hard disk loading, and (v) software counterfeiting. These forms are briefly discussed as follows:

9. Explain end user piracy

This occurs when a company employee reproduces copies of software without authorization. End user piracy can take the following forms:

1. Sing one licensed copy to install a program on multiple computers;
2. Copying disks for installation and distribution;
3. Taking advantage of upgrade offers without having a legal copy of the version to be upgraded;
4. Acquiring academic or other restricted or non-retail software without a license for commercial use; and
5. Swapping disks in or outside the workplace.

10. What is client-server overuse?

This type of piracy occurs when too many employees on a network are using a central copy of a program at the same time. If you have a local-area network and install programs on the server for several people to use, you have to be sure your license entitles you to do so. If you have more users than allowed by the license, that's "overuse".

11. Define internet piracy

This occurs when software is downloaded from the Internet The same purchasing rules should apply to online software purchase as for those bought in traditional ways. Internet piracy can take the following forms:

1. Pirate websites that make software available for free download or in exchange for loaded programmes;
2. Internet auction sites that offer counterfeit, out-of-channel, infringing copyright software; and
3. Peer-to-Peer networks that enable unauthorized transfer of copyrighted programs.

12. What do you mean by hard-disk loading?

This occurs when a business who sells new computers loads illegal copies of software onto the hard disks to make the purchase of the machines more attractive. The same concerns and issues apply to Value Added Resellers (VAR) that sell or install new software onto computers in the workplace.

13. What is software counterfeiting?

This occurs when pirates deliberately and illegally duplicate and sell copyrighted material, often making it appear to their customers that they are purchasing an authentic product. When it comes to packaged software, like that sold by Adobe, it is common for pirates to market and sell counterfeit copies of the software CDs along with reproduced packaging, manuals, license agreements, labels, registration cards, and security features.

Illegally copied and sold software is not eligible for support, training, or upgrades. You may not be able to register it, so it may not work properly. And if a vendor is willing to sell you something they don't have the right to sell, you have to wonder what else might get installed on your computer with the software you think you're getting.

14. What are the penalties against infringement of copyrights?

Penalties for copyright infringement can include:

(i) An injunction against further infringement such as an order preventing, the infriger from further copying or distribution of the copyrighted work.

(ii) Impounding or destruction of infringing copies.

(iii) Charging Damages: Either actual damages and the infringer's profit or statutory damages are charged.

(iv) Costs and attorney's fees

A copyright owner can only sue for infringement on a work that was registered with copyright office. The action against has to be taken within three months of first publication.

15. What are the penalties against infringement of patent, trademark and industrial designs?

The penalties against infringement of patent, trademark and industrial designs are as follows:

(i) **Patents:** A patent owner may recover money damages in the form of a reasonable royalty. If the infringement is found willful, three times money is recovered.

(ii) **Trademarks**: Same action as in case of patents can be taken against infringement of trademarks.

(iii) **Industrial Design**: The owner has the right to take legal action against the infringer and claim damages. However, the legal action must be taken within three years of the alleged offence.

16. What are the penalties against Software piracy?

There are significant financial and legal penalties for software piracy. Illegal distribution and use of software can result in prison terms of up to five years, and heavy fines [up to US$250,000 in USA]. In civil litigation against those who illegally distribute Adobe software, Adobe can obtain the higher of its lost profits, the infringer's profits, or statutory damages plus attorney fees. But the costs are not measured in money alone. When organizations are legally prosecuted for software piracy, they also suffer significant damage to their reputation.

Chapter 11

Domain Name

1. What is a domain name?

Domain refers to the unique name that identifies an Internet site. Domain names have two or more parts separated by dots. For example www.ask-edi.com or www.bcbsks.com. Also known as web address.

2. What criteria should be taken into account in choosing a domain name?

There are guiding lines which should be given due importance while selecting a domain name. Some tips for choosing a domain name are as follows.

(i) **Smaller Size:** The domain should not be too long. The domain name can be registered up to 63 characters long, but it is not necessary to use all 63 characters. In the case of domain names the smaller the better. Of course short domain names are harder to find but if you can find one it has a number of advantages such as, it's easier to brand, easier to remember and it's harder to misspell.

(ii) **Easy to remember:** Try to get a name that's easy to remember and sounds good. General words are best for this but since these are all gone at this stage you should look for non general names or combinations of general names which rhyme and sound good together. Take elephantdroppings.com for example that's an easy to remember name.

(iii) **Relate to your business:** If you use keywords which are related to your business industry you will help your search engine rankings. What's even better is if you can get your actual business name or a name similar to your company name as your domain name, this way all your off-line branding will pay off online as Internet users should be able to guess your domain name from your business name.

(iv) **Avoid similarities:** This means you should avoid choosing domain names that already have similar counterparts online, if you choose one of these people could end up on your competitors site rather than yours because they got confused over the two domain names.

(v) **.com is best:** You should always aim for the.com version of a name, that's what most people will go to after doing a search or after typing something directly into their address bar, if you register.tv or.net or any of the other available extensions and your competition has the.com version then you've just lost a potential customer.

(vi) **Use of Hyphen:** Use hyphens if your name is more than 3 words long (unless your name includes 3 short words). A name like carbuying.com doesn't really need a hyphen because it's readable the way it is. However a name like used-car-buying-scams.com looks better than usedcarbuyingscams.com since there are so many words in the name.

(vii) **Use of multiple Extensions:** Protect your name by registering multiple extensions (.net,.org, *etc.*) This deters people from copying your name. For example, if you own garden-tips-for-us.com you can also register garden-tips-for-us.net and any other extension that's available.

3. What other points should be considered in selecting a domain name?

The following points should also be kept in mind in choosing a domain name.

(i) Choose an appropriate top-level domain or domains.

(ii) Try for a variant of the name of your service or Web site.

(iii) Avoid names that are too narrow in scope (*e.g.*, the name of a short-term project or of one event).

(iv) Use Web-based wizards or other tools, such as DomainConquest, to look for unclaimed domain names that match key words.

(v) Avoid trademarks (even your own name, if it is someone else's trademark).

(vi) Avoid made-up words and alternate spellings (*e.g.*, "pythagoris") (they can be annoying and hard to remember), unless they are well established (*e.g.*, the name of your company).

(vii) Use at least one reasonably rare word (you have a better chance that the name is not already taken).

(viii) Avoid top-domain-like names (this can create problems with name resolution).

(ix) Don't use an embarrassing name or a name that may antagonize the wrong people.

(x) Don't use numeric digits at the beginning of the name (*e.g*, "525").

(xi) Don't use non-alphanumeric characters. Most of them don't work over the Internet. Hyphens and underscores are technically OK, but may be difficult for users to remember correctly. A name should never begin with "xn—", which is reserved for encoding Internationalized Domain Names.

(xii) Don't expect case to be preserved.

(xiii) For a large company, consider registering many variant forms and names of products, to avoid their use by competitors, critics, and purveyors of goods with which the company does not wish to be associated.

4. How will you register a domain name?

The process through which individuals and organizations obtain a domain name is called registration. Registration of a domain name enables the individual or organization to use that particular domain name for a specified period of time, provided certain conditions are met and payment for services is made. The important points about registration are listed below:

(i) Domain Names can be registered through many different companies known as registrars. The registrar you choose will ask you to provide various contact and technical information that makes up the registration. The registrar will then keep records of the contact information and submit the technical information to a central directory known as the "registry." This registry provides other computers on the Internet the information necessary to send you e-mail or to find your web site. You will also be required to enter a registration contract with the registrar, which sets forth the terms under which your registration is accepted and will be maintained.

(ii) The application is to be filed in a prescribed format called registration forms. These forms are used to register new domain names, new contacts for domain names, and new hosts (name servers) as well as to update domain name, contact, and host records.

(iii) The prescribed fee is to be deposited. It can be deposited by credit card, demand draft or in cash.

(iv) The domain name is allotted and activated after the payment of fee is received in the registrar office.

(v) Generic top level domains are available for registration on a global basis.

5. What is the registration Fee for a domain name?

It refers to the charge for registering, a domain name. The registration fee covers the cost of processing the initial registration and maintaining the domain name record for two years. The registration fee varies from country to country and company to company. After the initial two-year period, the domain name is subject to renewal (re-registration) and the renewal (re-registration) fee will be due on an annual basis if the registrant wishes to renew the domain name's registration.

6. What is the period of registration of a domain name?

A domain name is registered for a period of one year which has to be renewed every year. In other words, the domain name is renewed on yearly basis.

7. When does a domain name come into force?

The domain name becomes operative immediately after registration. It remains active upto 30 days after the expiry of the registration date.

8. Is it possible to transfer a domain name?

It refers to the process of changing the party whom is listed as the domain name registrant. The party taking over the domain name is responsible for paying a new registration fee. Domain name transfers are processed using the Registrant Name Change Agreement (RNCA).

9. What is infringement of a domain name?

The use of a domain name after the expiray of 30 days from the due date of registration amounts to infringement. It invites legal action against the unauthorized users.

10. What is a domain name system (DNS)?

A local database that translates domain names to Internet Protocol (IP) addresses is called domain name system. It is much easier for humans to remember and use names than IP numbers, which are what computers need to find each other on the Internet. Individual Internet computers around the globe maintain their specific portion of this database, and the data held in each portion of the database is made available to all computers and users on the Internet. The DNS comprises computers, data files, software, and people working together.

11. What are parts of a domain name?

Domain Names always have 2 or more parts separated by the dots. The left most piece is the most specific part while the right part is the TLD or Top Level Domain. Generally, there are three parts of a domain name such as (i) top level domain, (ii) second level domain, and (iii) third level domain.

12. What is the top level domain (TLD ?

The right most part of a domain name is called Top Level Domain TLD]. In other words, in a domain name, that portion of the domain name that appears in the right most position is known as top level domain. For example, the com in DomainMart.com. The main points related to TLD are given below:

(i) It is the right most part of domain name.
(ii) Every domain name ends with top level domain [TLD].
(iii) It is also known as first level domain.
(iv) It generally consists of three generic characters.

In the beginning there were only 5 Global top level domains [TLD's] that were recognized by everyone. These TLD's are given below:

- .com This TLD is intended for Commercial business around the world.
- .net This TLD is historically and is still used by Internet Service Providers.

- ☆ .org This TLD is intended to search the noncommercial community.
- ☆ .mil This TLD is restricted to use by the United States Military.
- ☆ .gov This TLD is restricted to use by United States Government Agencies.

13. What is second level domain?

In the Domain Name System (DNS), the next highest level to the left position of the top level domains is referred to as second level domain. In other words, it is that portion of the domain name that appears immediately to the left of the top level domain. For example, the Domain Mart in DomainMart.com.

14. What is third level domain?

In the Domain Name System (DNS, the next highest level of the hierarchy underneath the second level domains is known as third level domain. It is that portion of the domain name that appears two segments to the left of the top level domain. Third level domains are not the portion of an email address that appears in front of the @ symbol - for example, the webmaster in support@DomainMart.com is not a third level domain.

There can be **fourth and fifth level domains** and so on, with virtually no limitation. An example of a working domain with five levels is *www.sos.state.oh.us*. Each level is separated by a dot or period symbol between them.

Beside these, there are two more terms *viz.* generic top level domain, country code top level domain, subdomain and host nane. These are defined as follows:

15. What is generic top-level domain (gTLD)?

It is a top level domain of an Internet address used by a particular class of organization. These are three or more letters long, and are named for the type of organization that they represent. The generic top level domains include.com,.net,.org,.mil,.gov, *etc.* In addition, seven new gTLDs were also selected by ICANN (the Internet Corporation for Assigned Names and Numbers) on November 16, 2000. These are:.aero (for the entire aviation community);.biz (for business purposes);.coop (for cooperatives);.info (unrestricted);.museum (for museums);.name (for personal names);.pro (for professionals).

16. What is Country Code Top Level Domain [ccTLD] ?

A ccTLD is a country code top-level domain, for example:.mx for Mexico. These ccTLDs are administered independently by nationally designated registration authorities. There are currently 243 ccTLDs reflected in the database of the Internet Assigned Numbers Authority (IANA). WIPO, which has a ccTLD Program, has launched a database portal, facilitating online searches for information related to country code top level domains.

17. What is subdomain?

Generally known as a "Domain within a Domain", subdomains are web addresses built upon an existing domain name. In other words, a subdomain is a subdivision of a larger domain. For example, "mail.yahoo.com" is a subdomain of "yahoo.com."

18. What are hostnames

The hostname of an Internet Address is all of the left most pieces of a full internet address if there are more than 2 parts to the address. If there are only 2 parts of the address, then the hostname is equivalent to the domain name. Some examples are as follows:

Full Address	*Host Name*	*Domain Name*	*TLD*
www.keycomputer.com	www	keycomputer.com	com
keycomputer.com	keycomputer.com	keycomputer.com	com
www.google.net	www	google.net	net
my.www.keycomputer.com	my.www	keycomputer.com	com

It is clear from the above examples that the hostname can be many different things.

(i) If the entire address consists of only 2 parts, then the hostname is equivalent to the domain name.

(ii) If the entire address consists of more than 2 parts, then the hostname is equivalent to all the left most pieces before the right 2 parts (or the domain name)

In other words, a hostname refers to a domain name that has one or more associated Internet Protocol [IP] addresses. For example, the en.wikipedia.org and wikipedia.org domains are both hostnames, but the org domain is not.

19. What are Types of Domain Name Extensions ?

A domain name extension is also called a top level domain name such as.com,. org,.net, and a lot more. It is the top most part of the web site address. The domain name extensions are mainly of three types, *viz.*, (i) International domain name extension, (ii) National domain name extension, and (iii) alternative domain name extension.

20. What are International domain extensions?

International Domain Extensions are also referred to as common domain extensions. Three domain name extensions, *viz.*,.com,.net, and.org are the most common on the international front. Moreover, these domain names are called as the "Big 3" because they are accessible to anyone, without limitations. They are now widely recognized in all parts of the globe.

21. What are National domain extensions?

National Domain Extensions are also known as Country Code Domain Extensions. Now most countries have a matching top level domain name, under which domain names are sold. The most common domain name extensions are based on two characters ISO country codes. Most popular country top level domains include the.fr for France,.jp for Japan, and.co.uk. These domain name extensions are generally under the control of the government and have high level of stability and durability. The Country Code Top Level Domains [ccTLD] are becoming more popular due to the reducing supply of the more common.com,.net and.org extensions.

22. What are alternative domain extensions?

In the year 2000 the ICANN [Internet Corporation for Assigned Names and Numbers]approved a new set of seven top domains such as .aero, .biz, .coop, .info, .museum, .name, and .pro. The uses of some generic top level domains are explained as follows:

TABLE 11.1: Uses of some Generic Top Level Domains

Sl.No.	*Domain Extension*	*Meant for or designated for*
1.	.com	For the commercial organizations.
2	.net	For network services like Internet Service Providers.
3	.org	For non-profit organizations.
4	.aero	For entire aviation community
5	.biz	For business purposes
6	.coop	For cooperatives
7	.inf	For unrestricted use
8	.museum	For museums
9	.name	For personal names
10	.pro	For professionals

They are generally called "alternative top level domain names". They are officially known, thus will work in all parts of the globe, in any browser or other software. These are also known as generic top level domains.

Recently, ten new domain name extensions have been included. These are are. asia,.cat,.jobs,.mail,.mobi,.post,.tel,.travel, and.xxx.

23. What is Relative Significance of different domain extensions?

There are many types of domain extensions and each has its importance and value The importance of some domain extensions is given below:

(i) **".com"** for example is perceived in the public as more credible and usable than a ".tv" or ".mobi", it is more "official." In the academic world, the ".com", is preferred rather ".org" or ".edu".**com** These sites are generally businesses of some sort or informational portals. The most popular extension, this is the one people know about.

(ii) **.org, edu** These sites are more credible and usually often are dedicated to a school or non-profit organization.

(iii) **.tv, .mobi, .us,** *etc.* These extensions generally are different types of domain extensions associated with how you access them **.mobi**, for example, refers to accessing it through your mobile phone. They are not as hot as a**.com,** though still very popular and growing.

24. What the role of the internet corporation for assigned names and numbers [ICANN] ?

ICANN refers to the Internet Corporation for Assigned Names and Numbers. ICANN is responsible for managing and coordinating the Domain Name System (DNS) to ensure that every address is unique and that all users of the Internet can find all valid addresses. It does this by overseeing the distribution of unique IP addresses and domain names. It also ensures that each domain name maps to the correct IP address.

ICANN is also responsible for accrediting the domain name registrars. "Accredit" means to identify and set minimum standards for the performance of registration functions, to recognize persons or entities meeting those standards, and to enter into an accreditation agreement that sets forth the rules and procedures applicable to the provision of Registrar Services.

ICANN's role is very limited, and it is not responsible for many issues associated with the Internet, such as financial transactions, Internet content control, spam (unsolicited commercial email), Internet gambling, or data protection and privacy.

25. What are the differences between trademarks and domain names?

There are several differences between trademarks and domain names which presented in Table 11.2.

TABLE 11.2: Differences between Trademark and Domain Name

Sl.No.	*Particulars*	*Trademarks*	*Domain Names*
1	Registration with	Trademarks office	Registrar Internet
2	Validity	Within country	Both global and national
3	Items covered	Identification marks	Internet web site
4	Enforcement	With commercial use	After registration
5	Annual Fee	Charged	Charged once

26. What are the similarities between Trademarks and Domain Names ?

There are some similarities between trademarks and domain names which presented in Table 11.3.

TABLE 11.3: Similarities between Trademark and Domain Name

Sl.No.	*Particulars*	*Trademarks*	*Domain Names*
1	Registration	Essential	Essential
2	Types	Many	Many
3	Control	By owner	By owner

Sl.No.	Particulars	Trademarks	Domain Names
4	Registration fee	Payable	Payable
5	Renewal	Possible	Possible
6	Transfer	Possible	Possible
7	Requirement	Distinct	Distinct
8	Infringement	Possible	Possible

27. What are the advantages of domain names?

There are several advantages of domain names. Some advantages of domain names are as follows:

(i) **Naming Rights:** A domain name may only be registered once and for one buyer only. Since it is in a first come, first served basis, the first one to register the name shall have the right to use the name. It is a memorable website address. The owner has complete control over his website.

(ii) **Identity:** A domain name represents an identity which visitors always recognize. It is just like a brand name or a logo.

(iii) **Simplicity:** A domain name shortens the URL address of your blog and it is much easier to remember. It is easier for users to remember domain name than host-domain-path which is longer.

(iv) **Adds Credibility:** A domain name projects a professional image of a owner and thus adds credibility to the owner. In other words, a domain name looks more professional and does have more perceived credibility from all your visitors.

(v) **More Advertisers:** Some advertisers prefer a domain name than a sub domain name. With the use of a domain name, there is an expansion of opportunities for your blog. More advertisers, more tasks.

(vi) **More impressive:** It is very impressive (looks less like a personal site).

(vii) **Transportable:** It is transportable to another hosting service.

(viii) Customer service is readily available.

(ix) Domain name makes the keywords clear to search engines because the words are separated.

28. What are the disadvantages of domain names ?

There are some disadvantages of domain names which are presented below:

(i) It must be registered with appropriate service, normally for a small yearly fee.

(ii) Sometimes the name we want may already be taken; we may have to pay a lot (or sue) to obtain it.

(iii) One may lose the domain name unexpectedly because of registrar incompetence.

(iv) It requires proper maintenance of DNS records on a server.

Abbreviations

IDN: Internationalized domain name
URL: Uniform Resource Locator
DNS: Domain Name System
ICANN: Internet Corporation for Assigned Names and Numbers
gTLD: GenericTop Level Domain
SLD: Second Level Domain
FLD: First level Domain
TLD: Third Level Domain
TLD: Top Level Domain
IP: Internet Protocol IP
HTTP: Hypertext Ttransfer Protocol

Chapter 12

Trade Name

1. What a trade name?

A trade name, also known as a trading name or a business name, is the name which a business trades under for commercial purposes, although its registered, legal name, used for contracts and other formal situations, may be another. Pharmaceuticals also have trade names (*e.g.* "Aspirin"), often dissimilar to their chemical names ("acetylsalicylic acid"). Trading names are sometimes registered as trademarks or are regarded as brands. A trade name is the name under which a product is commercially known. The commercial name by which a chemical is known is called trade name. One chemical may have a variety of trade names depending on the manufacturers or distributors involved.

2. What are the main points related to trade names?

Main points related to trade name are given below:

(i) It is also known as business name, trading name, assumed name, brand name and corporate name.

(ii) It may or may not be registered.

(iii) It distinguishes a particular business from others.

(iv) It can be exclusive or non-exclusive.

(v) It is generally used on letter heads and bank accounts.

(vi) It is the name given to a particular substance by each company that manufactures it.

(vii) It is the name under which a company conducts its business.

(viii) It is s the business name of the person or organization making and/or selling a particular product. "General Motors, Inc" is the "trade name" of the company.

(ix) It is used by a company to describe and distinguish its brand of a generic product. Kleenex is a trade name for a brand of tissue; Xerox, a single brand of copier.

3. What are the points that should be given due importance while choosing a business name?

There are a number of considerations in choosing a business name, and securing it for use in a trading for a company, partnership or sole trader. A business name is the name by which a sole trader, partnership, limited liability partnership or a company conduct business. A business name must have distinctive, descriptive and legal elements and each element is considered as part of the whole name. The name should not be too general.These are briefly explained below:

(i) **Distinctive:** This is the element which is the unique identifier of the business. This element is the main identifier of business name. This is the part of the name that sets it apart from other businesses of the same type. Every business name must have a distinctive element. A word by itself may not be distinctive but it would be distinctive when used in combination with other words.

(ii) **Descriptive:** This is the element which describes the main aspect of the business.

(iii) **Legal:** This is the element when the name is used by a corporation. The legal element indicates the status of the organization as a corporate body.

(iv) **It should not be too General:** Names that are too general usually lack the distinctive element. They infringe on existing registered names, thus leading to confusion among the public. Names which are too general will also limit the number of names available to the public in the future.

Moreover, a business name should not suggest or imply a connection with any government or government department A business name must not suggest a connection with the Royal Family. The name of a society must be such that it will not be confused with existing names of societies, corporations, partnerships or trade names. When choosing a name for a society, all of the other naming guidelines should be followed.

4. What are different types of business ownerships?

There are three types of business ownership or business structures which are most commonly used. These are Sole Proprietorship, Partnership, and Company or Corporation.

5. Describe briefly the Sole Proprietorship.

This is the simplest form of business in which one person performs all the functions that are required for the successful operation of a business. The proprietor secures the capital, hires any employees, establishes and operates the business, assumes all the risks, accepts all profits or losses and pays all the taxes. The proprietor is said to be self-employed.

- ☆ **Taxation:** Any income or losses are claimed on the owner's personal tax return each year. Business deductions are permitted.
- ☆ **Liability:** Liability arising out of the business activity is borne entirely by the proprietor. All assets, whether business or personal, can be legally taken and used to discharge the liability.
- ☆ **Registration:** A sole proprietorship can be operated under the individual's name without the need for registration. If a business name is used such and company the registration is required.A business name registration is valid for three years from the date of registration. At that time, registration will expire and renewal of registration must be completed by the sole proprietor of the business name or the registration will be removed.
- ☆ **Duration:** The business ends with the death or incapacity of owner. Therefore, this type of business cannot be inherited. However, the individual assets can be used to carry on a business. The business name can be removed when a sole proprietor discontinues business.

6. What is a partnership type of business?

A partnership is a voluntary contract between two or more competent persons to place their money, efforts, labour and skill in business with the understanding that there shall be a proportional sharing of profits and losses between them. A partnership is governed by the *Partnership Act.* For every partnership, there should be a written partnership agreement between the partners setting out their respective rights, interests and responsibilities.

- ☆ **Taxation:** A partnership agreement can allocate the profits or losses in any ratio agreed to between the partners but if there is no agreement, the profits must be allocated equally. Business deductions are taken by the partnership before the income is distributed to the partners and claimed on their personal tax returns.
- ☆ **Liability:** Partners are personally liable for debts of the partnership and the full amount of these debts can be collected from one or more of the partners rather than the debt being equally shared. Partners can also be held liable for acts committed by one of their partners in the normal course of business.
- ☆ **Registration:** Within three months of the formation of a partnership, a declaration of the partnership must be registered under the *Partnership Act.* A computerized name search is made of the proposed partnership name prior to registration. A partnership registration is valid for three years from the date of registration. At that time, the registration will expire and a renewal of registration must be completed by the partners or the partnership registration will be removed.
- ☆ **Duration:** This business organization ends with death, incapacity, withdrawal or bankcorrupcy of any partner, unless otherwise agreed to in a Partnership Agreement. When a partnership discontinues business, the business name can be removed from the business names database.

7. Describe briefly corporation type of business.

A corporation or company is a separate legal entity having an existence separate and apart from the business person that created it. A company can be formed under the Company Act or the Corporations Act. The provincial corporations are most common. Its continuance depends only on the filing of annual returns and annual meetings of shareholders and directors. The Company Act allows a company to be owned by just one person. If more than one person is involved in the business, a Unanimous Shareholder's Agreement is recommended to establish the operational rules between the shareholders including dispute resolution and share evaluation. It is wise to consult with a lawyer to ensure that the company is set up in a manner that meets your business' present and future needs.

- ☆ **Taxation:** The Company pays taxes on corporate earnings after the business deductions are made. Money taken out of the company in the form of share dividends or wages by the share holders is claimed on their personal tax returns.
- ☆ **Liability:** Liability is limited to the assets of the company and the shareholders are not personally liable for the debts of the company unless allowed by law. The directors of a company may be held personally liable under certain circumstances, such as unpaid contributions under the Income Tax Act. Most banks and many other large creditors will require the owner of a small company to provide a personal guarantee to secure a loan. These factors reduce the effectiveness of limited liability for small business owners.
- ☆ **Registration:** A corporation can register a business name under the *Partnership Act* in the same manner that an individual(s) can register a Sole Proprietorship or Partnership. A corporation may wish to use a name, other than their corporate name, for the conduct of some or all of their business activity. If a business name is used by a corporation, a declaration of this in writing must be filed. An officer authorized by the corporation must complete the declaration on behalf of the corporation. A corporation wishing to register a trade name is required to have a computerized name search of the proposed name prior to registration.
- ☆ **Duration:** A business name registration is valid for three years from the date of registration. At that time, registration will expire and renewal of registration must be completed by an officer of the corporation or the business name will be removed. When a corporation discontinues the use of a trade name, the name can be removed from the business names database. The name can be removed on written request only by the registered owners of the business name.

  A company is its own legal person, therefore it survives the death, incapacity or withdrawal of any shareholder or director. Its survival depends on the filing of annual returns and fulfillment of the requirements of the Company Act.

8. What are the differences among sole proprietors, partner and corporation types of business structures?

A comparison of three types of business structures is presented in Table 12.1.

TABLE 12.1: Comparison of Three Forms of Business Structures, *viz.* Sole Proprietorship, Partnership and Corporation

Sl.No.	*Particulars*	*Sole Proprietors*	*Partners*	*Corporations*
1	Registration	Not required	Required	Required
2	Duration of registration	Three years if registered	Three years	Three years
3.	Renewal of registration	After 3 years	After 3 years	After 3 years
3	Liability	Owner	Partners	Share holders
4	Ownership	Individual	Partners	Share holders
5	Transfer	Permitted	Permitted	Permitted
6	Term terminates	With death of owner	After specified period	Unlimited
7	Infringement	Possible	Possible	Possible
8	Rights with	Owner	Partners	Board of Directors
9	Enforcement	With start of business	After registration	After registration
10	Validity	In country of registration	In country of registration	In country of registration

9. Is registration required for trade name?

It may or may not be registered. However, registration is necessary to get legal rights. Trade names are registered by the State in which the person or company is based. For each State, it has to be registered separately. The application is filed in the prescribed form and prescribed processing fee is to be deposited.

If a person (or persons associated in partnership) is engaged in business for trading, manufacturing, contracting, or mining purposes, he or she must register the business name by filing a declaration with the Registrar through an authorized service provider. The authorized service provider will examine your information and if it meets the current legislative requirements, will process the request and issue you a proof of filing to verify that the registration has occurred.

10. What are the matters covered by trade name?

Trade names cover all types of business, *viz.* sole proprietership, partnership; corporate business, *etc.*

11. What is the duration of protection for trade name?

Initially the trade name is registered for a term of three years which is renewable for a further term of three years each.

12. What are the Rights covered by trade name?

The sole owner of the trade name has all the rights related to his business, partners have equal rights and in a corporation, board of Directors have the rights.

13. What are the requirements for protection of trade name?

The basic requirements for registering a trade name are that that the name should be new and distinctive. It should not be confused with the trade name of any other person or corporation.

14. Is it possible to authorize or transfer trade name?

The owner of the trade name has right to transfer his business to any other person who is interested in the business. For transfer the request is submitted in the prescribed form and transfer fee is to be paid.

15. When does the trade name become effective?

The trade name comes into force immediately after registration and it expires after death of the owner or after its term if not renewed timely.

16. What is the Infringement of trade name?

Unauthorised use of a trade name by another without permission amounts to infringement. The trade name owner has rights to prevent such use. He can take legal action against infringer.

17. What are the differences between trademark and trade name?

There are some similarity and some differences between trademarks and trade names which are presented in Table 12.2.

TABLE 12.2: Comparison of Trademarks and Trade Names

Sl.No.	*Particulars*	*Trademarks*	*Trade Names*
1	Governed by	Primary Rights	Primary Rights
2.	Registration	Compulsory	Compulsory
3	Period of protection	Renewable indefinitely	Renewable
4.	Protection over	Names, images and short phrases	Products
5.	Validity	In original country	Country of registration
6.	Enforcement	With commercial use	With start of business
7.	Requirements	New and distinctive	New and Distinct
8.	Transfer	Permitted	Permitted
9	Infringement	Possible	Possible

A trade name identifies the business itself, whereas trademarks identify goods or services. A trade name can also serve as a trademark if it meets the requirements of a trademark.

18. What are advantages and disadvantages of trade names?

Advantages and disadvantages of sole proprietorship are briefly presented below:

Sl.No.	Advantages	Disadvantages
1	Low start-up costs	Lack of continuity
2	Greatest freedom from regulation	More difficult to raise capital
3	Direct control by owner	—
4	Minimum working capital requirements	—
5	Tax advantage to small owner	—
6	All profits to owner	—

19. What are advantages and disadvantages of partnership type of business?

Advantages and disadvantages of partnership are given below:

Sl.No.	Advantages	Disadvantages
1	Ease of formation	Unlimited personal liability
2	Low start-up costs	Lack of continuity
3	Additional sources of venture capital	Divided authority
4	Broader management	Difficulty in raising additional capital
5	Limited outside regulation	Hard to find suitable partners

20. What are advantages and disadvantages of corporation type of business structure?

The advantages and disadvantages of corporation type of business structure are presented below:

Sl.No.	Advantages	Disadvantages
1	Limited liability	Closely regulated
2	Specialized management	Most expensive to organize
3	Ownership is transferable	Charter restrictions
4	Continuous existence	Extensive record-keeping necessary
5	Legal entity	Double taxation, except when organized as an "S Corporation"
6	Easier to raise capital	Difficult to liquidate investment
7	Unity of action account having	
8	centralized authority in board of directors	

21. What are the differences between business names and corporation names?

There are differences between business names and incorporated names, as well as differences between the resulting entities formed. An incorporated name must contain a legal element, while a trade name or partnership name cannot. Another difference between a business name and an corporation name is that a corporation name is only granted once.

A difference between a registered corporation and a trade name or partnership, is that the trade name or partnership is not a distinct legal entity. Therefore, the proprietor(s) or partners are completely responsible for the debts and liabilities of the business.

Section II

***Sui-Generis* Rights**

Chapter 13

Intellectual Property Law: *Sui-Generis* Rights

1. What are *sui-geneis* rights?

Sui-generis is a Latin expression, literally meaning of its own kind/genus or unique in its characteristics. In law, it is term of art used to identify a legal classification that exists independently of other categorizations because of its uniqueness or due to the specific creation of an entitlement or obligation. In Intellectual Property there are rights which are known as being *sui-generic* to owners of a small class of works. Such Intellectual Property Rights include (i) database rights, (ii) mask work, (iii) plant breeders' rights, (iv) farmers' rights, (v) moral rights, (vi) supplementary protection certificate, and (vii) indigenous intellectual property.

2. What do you mean by database rights?

Database rights are a form of exclusive right introduced by European Union Law in 1996 in those countries which follow EU Law. A 'database right' is an intellectual property right given to a computer database. The main points about database rights are as follows:

(i) In most countries database are covered by copyright law to some degree.

(ii) The database rights were introduced by European Union Law to provide greater protection to collection of information.

(iii) Database right is covered by *sui-generic* rights.

(iv) **Duration:** Database rights last for 15 years, but can be extended if the database is updated.

(v) **Protection:** Database right prevents copying of substantial parts of database. The protection is over the information, not over the form of expression.

3. What is mask work right?

A mask work is a two or three-dimensional layout of an integrated circuit (IC or "chip"), *i.e.* the arrangement on a chip of semiconductor devices such as transistors and passive electronic components such as resistors and interconnections. Main points related to mask work are listed as under:

(i) Mask work can not be protected under copyright law or patent law.

(ii) The mask work exclusive rights were first granted in USA by Semiconductor Chip Protection Act of 1984. Such legislations also exist in Canada and Australia.

(iii) Duration. Rights for semiconductor mask work lasts only for two years (if unregistered) or 10 years (if registered).

(iv) The exclusive rights in mask work are somewhat like those of copyright: the right to reproduce the mask work or distribute an IC made using mask work.

(v) Reproduction for reverse engineering of a mask work is permitted by law.

(vi) Mask work rights exist when they are created, even without registration.

(vii) Protection under mask work cannot be provided to a mask that is not original.

4. What do you mean by plant breeders' rights?

Plant breeders' rights, also known as plant variety rights (PVR), are intellectual property rights granted to the breeder of a new variety of plant. Plant Breeders Rights are granted to novel plant varieties that are distinctive, uniform, and stable (eg, cultivars bred true-to-type for desired traits). The legal protection of a new plant variety is granted to the breeder or his successor. The effect of PBR is that prior authorization is required before the material can be used for commercial purposes.

5. What are advantages of PBR?

There are several advantages of plant breeders' rights which are listed below:

(i) Breeders get benefit of their variety.

(ii) PBR help in faster development of seed industry.

(iii) PBR lead to improvement in quality because of competition.

(iv) PBR are useful in procurement of good material on payment basis.

(v) PBR help in enrichment of genetic resources.

6. What are disadvantages of PBR?

There are several advantages of plant breeders' rights which are listed below:

(i) It will promote monopoly.

(ii) It will encourage unhealthy practices.

(iii) It may lead to increase in prices.

(iv) There will be reduction in genetic variability.

(v) There will be compulsion to purchase frsh seed every year.

7. What are Farmers' Rights?

Farmers' rights refer to the rights arising from the past, present and future contributions of farmers in conserving, improving and making available plant or animal genetic resources, particularly those in the centres of origin/diversity. In other words, the legal rights provided to farmers to save, use, sow, replant, exchange, share or sell his farm produce including seed of a variety protected under Plant Variety Protection Act refer to Farmers Rights. The purpose of these rights is to "ensure full benefits to farmers and support the continuation of their contributions.

The FAO Conference held in Rome from 11-29 November, 1989 endorsed the concept of Farmers Rights with a view to:

(i) Ensuring global recognition of the need for conservation and the availability of sufficient funds for these purposes;

(ii) Assisting farmers and farming communities throughout the world, especially those in areas of original diversity of plant genetic resources, in the protection and conservation of their PGR and of the natural biosphere; and,

(iii) Allowing the full participation of farmers, their communities and countries in the benefits derived, at present and in the future, from the improved use of PGR.

8. What are moral rights?

Moral rights are rights of creators of copyrighted works generally recognized in civil law jurisdictions and first recognized in France and Germany, before they were included in the Berne Convention for the Protection of Literary and Artistic Works in 1928. Moral rights are a special extension to copyright that give the originator of a copyrighted work rights over its use. The originator can specify moral conditions on the use or exploitation of their creation, even when the rights are licensed or sold to others'.

9. Define supplementary protection certificate.

This term is used in European Countries. This is an extension of the term given to pharmaceutical or plant protection patent. A supplementary protection certificate (SPC) is a *sui-generis,* patent-like, intellectual property right. This type of right is available for medicinal products, such as drugs, and plant protection products, such as insecticides, and herbicides.

10. What do you mean by traditional knowledge?

The Indigenous Knowledge also known as Traditional Knowledge [TK] refers to the knowledge, innovations and practices of indigenous people and local communities. In other words, it is the: knowledge gained through tradition from ancestors about plants and animals, medicines, agricultural practices and other matters.

11. Define ecological knowledge.

The ecological knowledge is composed of both traditional knowledge and experiential knowledge[gained through personal experience. Traditional knowledge is very important almost in every sphere of life. Hence, its protection is very necessary otherwise this valuable knowledge will be gradually lost.

12. What are differences between Database rights and copyrights?

There are some differences between database rights and copyrights which are presented in Table 13.1.

TABLE 13.1: Differences between Database Rights and Copyrights

Sl.No.	*Particulars*	*Database Rights*	*Copyrights*
1	Governed by	*Sui-generis* rights	Primary intellectual rights
2	Period of protection	15 years	60-110 years
3	Protection over	Information	Expression

13. What are the similarities between database rights and copyrights?

There are some similarities between database rights and copyrights which are presented in Table 13.2

TABLE 13.2: Similarities between Database Rights and Copyrights

Sl.No.	*Particulars*	*Database Rights*	*Copyrights*
1.	Registration	Not compulsory	Not compulsory
2	Validity	Country of registration	Country of registration
3	Enforcement	After creation	After creation

14. What are differences between database rights and mask work rights?

There are some differences between database and mask work which are presented in Table 13.3.

TABLE 13.3: Differences between Database and Mask Work

Sl.No.	*Particulars*	*Database Rights*	*Mask Work*
1	Period of protection	15 years	Unregistered 2 Registered 10 years
2	Protection over	Information	Integrated Circuits

15. What are similarities between database rights and mask work rights?

There are some similarities between database and mask work which are presented in Table 13.4.

TABLE 13.4: Similarities between Database and Mask Work

Sl.No.	*Particulars*	*Database Rights*	*Mask Work*
1	Governed by	*Sui-generis* rights	*Sui-generis* rights
2.	Registration	Not compulsory	Not compulsory
3	Validity	Country of registration	Country of registration
4	Enforcement	After creation	After creation

Chapter 14

Database Rights

1. What are database rights?

The database refers to a collection of data organized for rapid search and retrieval by a computer. In other words, database refers to an organized collection of information, data, or citations stored in electronic format that can be searched for specific information or records by techniques specific to each database. A 'database right' is an intellectual property right given to a computer database. A database is defined as a collection of independent works, data or other materials which: (a) are arranged in a systematic or methodical way; and (b) are individually accessible by electronic or other means.

2. Highlight main point related to database rights.

The database rights are of recent origin. The database rights were first introduced by European Union Law 1996. In U.K. it was introduced in 1997 but came into force from 1st January 1998. The purpose of this new database right is to protect the investment in a database. A database right arises automatically at no cost to you. Other important points about database rights are as follows:

(i) Database rights were first adopted by U.K. in 1998.

(ii) Database rights are covered by *sui-generic* rights.

(iii) In most countries database rights are covered by copyrights.

(iv) Database rights exist in most European countries but not in the rest of the world.

(v) Members of WTO have an obligation to provide copyright protection for some databases.

(vi) A database may be protected by copyright and/or the database right.

(vii) The protected database is permitted for normal usage [see later].

3. Is registration compulsory for database rights?

In many respects, a database right is similar to copyright. It is created automatically. In other words, this is an automatic right which protects the investment (*i.e.* time, money and energy) that goes into obtaining, verifying or presenting the contents of a database. There is no need of registration of database. But it is advisable to register the databse, because registration provides legal rights against unauthorised use of database by other persons. It is a right against use and not a monopoly.

4. What are the requirements for protection of database rights?

The owner of a protected database is a person who has taken the risk of investing in the creation of database. For the database rights to apply, the database must be author's own intellectual creation and the result of substantial investment (financial, technical or otherwise). A database will qualify for protection if it meets the following requirements:

(i) **Author's own Creation:** The database is a "collection of individual works, data or other materials arranged in a systematic or methodical way", and

(ii) **A substantial investment:** There should be a proof that substantial investment has been made in creation and maintenance of database.

5. What types of matters are covered by database rights?

Any work that shows originality in its selection, coordination and arrangement can be protected under database rights. The following items are covered by database rights:

(i) Telephone directories or telephone book,

(ii) Railway tome tables,

(iii) Electronic photo book,

(iv) Encyclopedia with different entries, and

(v) Microfilm collections, *etc.*

6. What are the items that are not covered by database rights?

The following items are not covered by database rights:

(a) Any computer software used to access or maintain the database.

(b) A movie, a recording, an audiovisual, cinematographic, literary or musical work.

7. What is the Duration of Protection for database rights?

The term for protection of database rights is 15 years from the end of the year in which the database was completed or first made avilable to the public.The term is extended automatically, if a substantial change [additions, deletions or alterations] is made to the datbase during the term of protection The changes include substantial alterations, deletions and additions to the database in such a way so as to reflect substantial new investment.

8. What are the Rights Provided to database owner?

A database right prevents copying of substantial parts of a database (including frequent extraction of insubstantial parts). However, unlike copyright the cover is not over the form of expression of information but of the information itself. The database owner is entitled to prevent others from:

(i) Extraction and or re-utilization of the whole or of a substantial part, evaluated qualitatively and/or quantitatively, of the contents of that database'.

(ii) Repeated and systematic extraction and/or re-utilization of insubstantial parts of the contents of the database implying acts which conflict with a normal exploitation of that database or which unreasonably prejudice the legitimate interests of the maker of the database'.

9. Is authorization of database possible?

The owner of the database has the rights to authorize other person to use his database on the basis of royalty payment. In other words, the holder of database has right to sale or licence his database.

10. When does the protection of database become effective?

The database rights come into force from the end of the year in which the the making of database was completed or the database was made available to the public.

11. How does the Infringement of database occur?

The rights holders may take action if their database is copied or used without permission. Database right can be infringed by:

(i) Copying the contents of a database and rearranging them.

(ii) Making the contents of the database available to the public without permission.

(iii) Unauthorized repeated and systematic extraction of 'insubstantial parts' of database.

(iv) Permanent or temporary transfer of the database contents to another medium by any means, or in any form.

As with copyright, if a database right is infringed a range of civil remedies are available to the database right holder including injunctions, an account of profits, damages and delivering up of goods.

12. What is the normal usage of database?

The extraction or reutilization of substantial part of a published data is permitted without permission of the database creator under following circumstances:

(i) If the data is used for teaching or research and not for any commercial purpose.
(ii) When the identity of the data creator is not ascertainable or
(iii) When the term of database right is expired.

13. What are differences and similarities between database rights and copyrights?

Database right is in many ways very similar to copyright so that, for example, there is no registration for database right – it is an automatic right and commences as soon as the material that can be protected exists in a recorded form. Ownership of database rights is determined in the same way as copyright ownership. As with copyright, database rights can be sold or licensed. The database right can apply to both paper and electronic databases. However, there are some differences between database right and copyright:

1 The term of protection for database right is much shorter. Database right lasts for 15 years from publication. It is possible that the right will last in perpetuity as the 15 year period of protection rolls forward if substantial changes and updates or other sufficient investments are made to the database in this period.

2. The activities that a database right holder can control, and which are, therefore, infringed if undertaken without the right holder's permission, are a bit different. Database right concerns control over the extraction and re-utilization of the contents of the database.

TABLE 14.1: Comparison of Database and Copyrights

Sl.No.	*Particulars*	*Database Rights*	*Copyrights*
1	Governed by	*Sui-generic* rights	Primary intellectual rights
2.	Registration	Not compulsory	Not compulsory
3	Period of protection	15 years	70-120 years
4.	Protection over	Information	Expression
5.	Validity	Country of registration	Country of registration
6.	Enforcement	After end of the year of creation	After creation

14. What are the advantages of database rights?

The Database Management System [DBMS] is a central system which provides a common interface between the data and the various front-end programs in the application. It also provides a central location for the whole data in the application to reside. The DBMS has several advantages over file based database management system as discussed below:

(i) **Minimal Data Redundancy:** Since the whole data resides in one central database, the various programs in the application can access data in different data files. Hence data present in one file need not be duplicated in another. This reduces data redundancy. However, this does not mean all redundancy can be eliminated. There could be business or technical reasons for having some amount of redundancy. Any such redundancy should be carefully controlled and the DBMS should be aware of it.

(ii) **Data Consistency:** Reduced data redundancy leads to better data consistency.

(iii) **Data Integration:** Since related data is stored in one single database, enforcing data integrity is much easier. Moreover, the functions in the DBMS can be used to enforce the integrity rules with minimum programming in the application programs.

(iv) **Data Sharing:** Related data can be shared across programs since the data is stored in a centralized manner. Even new applications can be developed to operate against the same data.

(v) **Enforcement of Standards:** Enforcing standards in the organization and structure of data files is required and also easy in a Database System, since it is one single set of programs which is always interacting with the data files.

(vi) **Application Development Ease:** The application programmer need not build the functions for handling issues like concurrent access, security, data integrity, *etc.* The programmer only needs to implement the application business rules. This brings in application development ease. Adding additional functional modules is also easier than in file-based systems.

(vii) **Better Controls:** Better controls can be achieved due to the centralized nature of the system.

(viii) **Data Independence:** Data Independence can be defined as immunity of applications to change in physical representation and access technique. The provision of data independence is a major objective for database systems.

(ix) **Reduced Maintenance:** Maintenance is less and easy, again, due to the centralized nature of the system.

The architecture of the DBMS can be viewed as a 3-level system comprising the following:

(i) The internal or the physical level where the data resides,

(ii) The conceptual level which is the level of the DBMS functions, and

(iii) The external level which is the level of the application programs or the end user.

15. What are the disadvantages of database rights?

There are some disadvantages of database rights which are listed below:

(i) Database systems are complex, difficult, and time-consuming to design substantial hardware and software start-up costs.

(ii) Damage to database affects virtually all applications programmes.

(iii) Extensive conversion costs in moving form a file-based system to a database system.

(iv) Initial training required for all programmers and users.

Chapter 15

Mask Work Rights

1. What is a mask work?

A **mask work** is defined as "a series of related images, however fixed or encoded, having or representing the predetermined, three-dimensional pattern of metallic, insulating, or semiconductor material present or removed from the layers of a semiconductor chip product, and in which the relation of the images to one another is such that each image has the pattern of the surface of one form of the semiconductor chip product.

Mask work exclusive rights were first granted in the US by the Semiconductor Chip Protection Act [SCPA] of 1984. In Canada these rights are protected under the Integrated Circuit Topography Act 1990. Equivalent legislation exists in Australia and Hong Kong.

2. Discuss main points related to mask work.

Important points related to mask works are given below:

(i) Mask work protection is protected by ***sui-generis*** rights. These rights were created to protect specific rights where other (more general) laws were inadequate or inappropriate. The mask works are not clearly protectable under copyrights as well as patent law. Hence, they are protected for a limited period under *sui-generis* rights.

(ii) The correct symbol used in a mask work's copyright notice is ? (**M** enclosed in a circle).

(iii) Mask work rights are closer to copyrights than with other exclusive rights such as patents or trademarks.

(iv) They are used alongside copyright to protect a read-only memory (ROM) component that is encoded to contain computer software.

(v) Mask works are neither subject to any statutory fair use defense, nor the typical backup copy exemptions that provides for computer software.

3. Is registration compulsory for a mask work?

The registration is not compulsory for mask works. But registration is essential to get legal rights.

4. What is the procedure of registering a mask work?

The registration process consists of the following steps:

(i) **Filing Application:** The application has to be filed in the Patents and Copyrights office in the prescribed form which cab be obtained from the copyrights office. The application for registration may be submitted by the owner of the mask work or an agent duly authorized by the owner.

(ii) **Fee:** The prescribed processing fee has to be deposited alongwith application form.

(iii) **Declaration Certificate:** A declaration has to be submitted stating that the applicant is authorized to submit the application and the statements made are correct to the best of knowledge of the person filing the application.

(iv) The application should bear name and signature [original] of the applicant.

(v) For the same version of a mask work only one registration can be made.

Mask works must be registered within 2 years of first commercial exploitation. The date of registration is the date on which a complete application, fee, and an appropriate deposit representing an acceptable filing are filed with the Copyright Office. If the Mask Work is not registered within this time period, protection under the SCPA will be lost.

5. What are the matters covered by mask work?

Mask work rights cover two or three-dimensional layout or topography of an integrated circuit (IC or "chip"), *i.e.* the arrangement on a chip of semiconductor devices such as transistors and passive electronic components such as resistors and interconnections.

6. What is the duration of protection for a mask work?

The rights in semiconductor mask work last only two years (if unregistered) or ten years (if registered). The SCPA provides protection of a Mask Work for a period of 10 years from the date of first commercial exploitation, or the date of registration with the Copyright Office, whichever occurs first.

7. What are the rights covered by mask work?

During the term of protection, the owner of a mask work has the following exclusive rights:

(i) To reproduce the mask work by optical, electronic, or any other means;

(ii) To import or distribute a semiconductor chip product in which the mask work is embodied; and

(iii) To induce or knowingly to cause another person to do any of the acts described in number (i) and number (ii).

The exclusive rights in a mask work are somewhat like those of copyright: the right to reproduce the mask work or (initially) distribute an IC made using the mask work. A owner of a mask work may freely import, distribute or use, but not reproduce the chip (or the mask).

8. What are the requirements for protection of mask work?

The legal protection is provided to the three-dimensional images or patterns formed on or in the layers of metallic, insulating, or semiconductor material and fixed in a semiconductor chip product, *i.e.*, the "topography" of the "chip." For protection, a mask work must be original. A mask work is considered "original" if it is the independent creation of an author who did not copy it.

9. Is authorization or transfer of a mask work possible?

The Act allows the owner of the exclusive rights to transfer all of them or to license all or fewer than all the rights. A transfer or license must be in writing and signed by the owner of the rights or by a duly authorized agent of the owner. The Act also provides that the exclusive rights in a mask work may be transferred by operation of law or by terms of a will; or, the rights may pass as personal property under the applicable state Law. In other words, the owner of a mask work can transfer all the exclusive rights of his mask work to any other person.The new owner should obtained by transfer the totality of the rights in the mask work.

10. When does the mask work come into force?

Protection for a mask work begins on the date the mask work is registered with the Copyright Office, or the date the mask work is first commercially exploited anywhere in the world, whichever occurs first. Protection lasts for 10 years.The protection expires at the end of the tenth calendar year after it began.

The mask work rights come into force immediately after creation of mask work. In other words, mask work rights exist automatically [like copyrights] when they are created, regardless of registration. In case of patents, rights are confered only after registration.

11. What is the Infringement of a mask work?

After a certificate has been issued by the Copyright Office, the owner of the mask work or the exclusive licensee of all rights in the mask work may institute a civil action for infringement occurring after the commencement of protection accorded by the Act.

12. What is the notice and symbol of a mask work?

An appropriate notice on a semiconductor chip or product includes either the words "Mask Work," the symbol *M*, or the symbol Ⓜ, followed by the name of the owner or a recognizable abbreviation of the owner's name. Such notice is optional, but provides certain benefits to the owner with respect to enforcing rights.

13. What are the activities permitted for a mask work?

It is important to understand that the Semiconductor Chip Protection Act [SCPA] does not provide as broad protection as the patent laws do. Certain activities by competitors are permitted, such as reverse engineering for the purposes of teaching, analyzing, or evaluating the concepts embodied in the mask work including the circuitry, logic flow, and organization of components. The results of legitimate reverse engineering may be incorporated into another original mask work and commercially exploited.

Any idea or concept, process, principle, discovery or system associated with [a semiconductor product] a mask work is not protected.

14. What are the similarities and dissimilarities between mask work and database rights?

There are some similarity and some differences between mask work and database rights which are presented in Table 15.1.

TABLE 15.1: Comparison of Mask work and Database Rights

Sl.No.	*Particulars*	*Mark Work*	*Database Rights*
1	Governed by	*Sui-generis* rights	*Sui-generis* rights
2	Registration	Not compulsory	Not compulsory
3	Period of protection	Unregistered 2 years Registered 10 years	15 years
4	Protection over	Integrated Circuits	Information
5	Validity	Country of registration	Country of registration
6	Enforcement	After creation	After the end of the year of completion or creation

15. What are the advantages of mask work rights?

There are several advantages of mask work which are listed below:

(i) Registration is optional.
(ii) It protects original work.
(iii) It promotes creative work.
(iv) It provides benefits to the owner.

16. What are the disadvantages of mask work rights?

There are some disadvantages of mask work which are listed below

(i) Protection is for shorter period than copyrights.
(ii) It does not provide legal rights without registration.
(iii) Rights are valid only in the country where registered.
(iv) It may lead to monopoly.

Chapter 16

Plant Breeders' Rights

1. What are plant breeders' rights?

The Plant Breeders' Rights programme was first established in 1987 under the Plant Variety Rights Act [PVRA], which was succeeded by the current Plant Breeders' Rights Act 1994. {PBRA].Plant breeders' rights [PBR], also known as plant variety rights (PVR), are intellectual property rights granted to the breeder of a new variety of plant. Plant Breeders Rights are granted to novel plant varieties that are distinctive, uniform, and stable (e.g, cultivars breed true-to-type for desired traits). The legal protection of a new plant variety is granted to the breeder or his successor. The effect of PBR is that prior authorization is required before the material can be used for commercial purposes.

2. What is the procedure for registration of plant breeders' rights?

Registration is essential to get legal rights. There is no protection without registration. The registration consists of following steps:

(i) Filing application in the prescribed form.

(ii) Payment of prescribed processing fee along with application. The registration fee varies from country to country. In USA, after registration annual renewal fee [300 Dollars] is payable.

(iii) Examination of the application by the experts.

(iv) Issue of protection certificate, if application if found to meet the desired requirements.

(v) Normally the registration takes about 2.5 years for most of the species.

3. What is the duration for protection of plant breeders' rights?

The period of protection varies with plant species. The protection is provided for a definite period. The duration of protection for different plant species is as follows:

(i) **For Trees and Vines:** Initially for nine years and maximum for 18 years.

(ii) **For extant varieties:** Initially six years and maximum for 15 years from the date registration of the variety.

(iii) **In other cases:** Initially six years and maximum for 15 years from the date registration of the variety.

Thus breeders' rights allow plant breeders to get benefit of their variety for a limited period. When the PBR period is over, the variety reverts to the public domain and is available to every body. The limited duration of Plant Breeders' Rights ensures a balance between private and public interest.

4. What is the validity of PBRs?

The protection of the Plant Breeders' Rights is valid only in the country where it has been registered. The protection in other countries can be obtained by filling separate application in each country.

5. What are the matters covered by plant breeders' rights?

The protection right can be granted for varieties of all botanical genera and species. The variety should have a designation [name] as per the rule of International Code of Nomenclature.

6. What are the requirements for protection of plant breeders' right?

There are four basic requirements for protection of a variety under PBR, *viz.*, (i) novelty, (ii) distinctiveness, (iii) uniformity, and (iv) stability. See later.

7. Is it possible to transfer plant breeders' rights?

An authorized plant breeder has right to authorize other interested persons for commercial production and marketing of his variety on royalty basis.

8. Who has control over the plant breeders' rights?

The authorized plant breeder has right to prevent other from commercial production and marketing of his variety without permission

9. When does the plant breeders' rights become effective?

The Plant Breeders' Rights come into force immediately after registration of the variety.

10. To whom plant breeders rights are granted?

BPR can be granted only to breeder or legal owner of a new variety.

11. What is protected by plant breeders' rights ?

The plant breeders rights protect the variety but not the standard breeding procedures that are used to develop new variety.

12. What is the purpose of plant breeders' rights?

The main purpose of PBR is to encourage breeders to develop better plant varieties. The PBRs provide monopoly of new varieties to breeders for a limited period.

13. Can the same variety be patented and protected by plant breeders' rights?

Yes, the same variety can be patented and protected by PBR. The gene technology or gene sequences can be patented and the end product that is the plant variety can be protected under PBR.

14. What are the requirements for protection of plant breeders' rights?

The PBR cannot be granted to any variety. There are four basic requirements for protection of a variety under Plant Breeders' Rights. These are (i) novelty, (ii) distinctiveness, (iii) uniformity, and (iv) stability.

15. What do you mean by novelty?

It refers to newness of a variety. The variety should be new and should not have been commercially cultivated for more than one year before granting protection under PBR. In other words, there should not be a prior sale for more than one year.

16. What is distinctiveness?

The new variety must be distinguishable in one or more morphological, quality or other character from previously available varieties. When a variety differs at least in one characteristic from varieties of common knowledge, it is known as distinct.

17. Define uniformity.

The variety should be uniform. In other words, all plants in a variety should look alike.

18. What is stability?

The should give stable performance in different generations. All these attributes are determined by grow out test.

19. What are the rights provided to the breeder of a new variety?

The *Plant Breeders' Rights Act* provides plant breeders the exclusive right to produce and sell new plant varieties which they have developed. In other words. It provides exclusive rights to the breeder for commercial production and marketing of his variety. The important Plant Breeders' Rights are as follows:

- **(i) Rights for commercial Seed production**: PBR provide legal right to the breeder of a variety for large scale seed production. This he can do either on his own farm or on the farms of authorized farmers on payment basis.
- **(ii) Rights for Marketing**: The breeder or owner of a variety has exclusive rights to regulate marketing of his variety.
- **(iii) Rights to export and import**: The breeder or owner of a variety has full rights to regulate export and import of his or her variety.

(iv) **Rights of Authorization:** The breeder or owner of a protected variety has rights to authorize other interested persons for commercial production, and marketing, export and import of his variety. However, prior authorization of the breeder or owner of a variety is required for such purpose.

(v) **Rights to prevent Infringement:** The breeder or owner of a variety has rights to prevent others from unauthorized commercial production and marketing of his variety.

20. What is infringement of plant breeders' rights?

Unauthorized production and marketing of a registered variety by other person amounts to infringement. The owner has the right to take legal action against the infringer and claim damages. The PBR Authority may initiate legal action against a person who is involved in the infringement of a protected plant variety. The Authority can recover both damages and profits from such person. The Plant Breeders' Act [PBRA] provides for heavy penalties against infringement of the breeders' right. In USA it is 55,000 Dollars for individuals and 2,75,000 Dollars for companies.

21. What are exemptions of plant breeders' rights act?

The Plant Breeders' Rights Act provides two exemptions, *viz.*, (i) breeders' exemptions and (ii) farmers' exemptions.

22. What do you mean by breeders' exemptions ?

The legal rights that are provided to plant breeders to use protected material for further research refer to breeders' exemptions. Breeders exemptions are also called research exemptions or breeders' privilege. The UPOV Act 1978 provides Breeders' exemptions. However, the Act 1991 has curtailed Breeders' Exemptions.

23. What are farmers' exemptions?

The legal rights that are provided to farmers to save, use, exchange, share or sell his farm produce of a protected variety are known as farmers' exemptions. Farmers' exemptions are also called Farmers' Rights or Farmers privilege. Here the sale is restricted to non-commercial sale. The UPOV Act 1978 provided farmers exemptions, however, these exemptions were curtailed by UPOV Act 1991.Because of these reasons, UPOV Act is not accepted by many countries.

24. What are advantages of plant breeders' rights?

There are several advantages of plant breeders' rights which are listed below:

(i) Breeders get benefit of their variety.

(ii) PBR help in faster development of seed industry.

(iii) PBR lead to improvement in quality because of competition.

(iv) PBR are useful in procurement of good material on payment basis.

(v) PBR help in enrichment of genetic resources.

25. What are disadvantages of plant breeders rights?

There are some disadvantages of plant breeders' rights which are listed below:

(i) It will promote monopoly.

(ii) It will encourage unhealthy practices.

(iii) It may lead to increase in prices.

(iv) There will be reduction in genetic variability

(v) There will be compulsion to purchase fresh seed every year.

Chapter 17

Farmers' Rights

1. What are farmers' rights?

Farmers' rights refer to the rights arising from the past, present and future contributions of farmers in conserving, improving and making available plant or animal genetic resources, particularly those in the centres of origin/diversity. In other words, the legal rights provided to farmers to save, use, sow, replant, exchange, share or sell his farm produce including seed of a variety protected under Plant Variety Protection Act refer to Farmers Rights. The purpose of these rights is to "ensure full benefits to farmers and support the continuation of their contributions.

The FAO Conference held in Rome from 11-29 November, 1989 endorsed the concept of Farmers Rights with a view to:

(i) Ensuring global recognition of the need for conservation and the availability of sufficient funds for these purposes;

(ii) Assisting farmers and farming communities throughout the world, especially those in areas of original diversity of plant genetic resources, in the protection and conservation of their PGR and of the natural biosphere; and,

(iii) Allowing the full participation of farmers, their communities and countries in the benefits derived, at present and in the future, from the improved use of PGR.

2. Is registration required for a farmers' Variety?

A farmer who has bred or developed a new variety shall be entitled for registration and protection of his variety like a breeder.

3. What is the procedure of registration of farmers' variety?

The registration process consists of the following steps:

(i) **Filing of application:** The application is to be filed in the prescribed form and submitted in the office of the registrar of Plant Variety Protection and farmers' Rights Act. The application should contain all the desired documents and information.

(ii) **Examination of the application:** The application is examined by experts in the registrar office. If found correct, it is advertised for opposition. Three months time is given for filing opposition.

(iii) **Registration:** If there is no opposition, and the registrar and Authority are satisfied, the variety is registered and the registration certificate is issued to the breeder, farmer or other owner of the variety.

4. What are the requirements for registration of farmers' variety?

The farmers; variety shall be entitled for registration if it meets the following requirements:

(i) It should have a denomination [name] assigned by the applicant.

(ii) An affidavit that the variety does not contain terminator gene.

(iii) A complete passport data of parental lines from which the variety has been derived.

(iv) The contribution of any farmer, village community, institution or organization in the development of the variety.

(v) The source of genetic material used in developing the variety and the parental material used in developing new variety was lawfully acquired.

(vi) A brief description of the variety about its novelty, distinctiveness, uniformity and stability.

(vii) A proof of the right to make the application.

5. How much fee is charged from farmers for registration of a variety?

In India, a farmer or group of farmers or village community shall not be liable to pay any fee in any proceeding before the Authority or Registrar or the Tribunal or the High Court under Plant Variety Protection and Farmers' Rights Act or the rules made there under.

6. What are the rights granted to farmer ?

The Farmers' Rights Act provide following rights to the farmers in India even for the variety protected under PVP and FR Act.

(i) Rights to save, use, sow or replant seeds of his farm produce.

(ii) Rights to exchange or share his farm produce, and

(iii) Rights to sell seeds of his farm produce.

However, the farmer cannot sell the seeds or planting material under a brand name.

7. What are rights of communities in India?

In India, any of the following can file application for registration on behalf of any village or community.

(i) Any person or group of persons [whether actively engaged in farming or not], and

(ii) Any governmental or non-governmental organization

After completing all formalities, if the Authority is satisfied, the variety can be registered. If due credit has not been given to the farmers or communities who have helped in the development of the variety, the breeder has to pay compensation to the concerned party, or part of compensation will be deposited in the gene fund.

8. Is authorization of farmers' variety possible?

The authorization for an essentially derived variety from farmers' variety cannot be given by the breeder of such farmers; variety except with the consent of the farmers or group of farmers or community of farmers who have made contribution in the preservation or development of such variety.

9. What are the materials covered by farmers' rights?

Farmers' Rights include new varieties of all self pollinated, cross pollinated and asexually propagated varieties.

10. What is the duration for farmers' variety?

The protection is provided for a definite period. The duration of protection for different plant species is as follows:

(i) For Trees and Vines: Initially for nine years and maximum for 18 years.

(ii) For extant varieties: Initially six years and maximum for 15 years from the date registration of the variety.

(iii) In other cases: Initially six years and maximum for 15 years from the date registration of the variety.

11. What is the validity of protection of farmers' rights?

The protection is valid only in the country where the variety has been registered.

12. What is benefit sharing?

If the variety has been developed by a particular farmer, he can get benefits of his variety. In case of community rights, the sharing of benefits becomes difficult and the money is deposited in the gene fund.

13. What is Gene Fund?

The central government may constitute a Fund to be called National Gene Fund. The money earned from following will be deposited in this fund:

(i) The benefit sharing received from the breeder of a variety.

(ii) The annual payable to the Authority in the form of royalty.

(iii) The amount of compensation received by the Authority.

(iv) Contribution received from amy National or International Organization and other sources.

The gene fund may be utilized for benefit sharing, rewards and other expenditure related to Plant Variety Protection and Farmers' Rights Act.

14. Who will be entitled for rewards?

A farmer who is engaged in the conservation of genetic resources of land races and wild relatives of economic plants and their improvement through selection shall be entitled for recognition and reward. The expenses toward rewards will be met out from the National Gene Fund.

15. What do you mean by acknowledgement by breeders?

It is compulsory for the breeder of a variety to reveal the origin of the parental lines and plant genetic resources used in the development of a variety. It will help in deciding the share of different persons, *viz.*, farmers and other communities in benefits earned by breeder of such variety.

16. Is infringement is exempt for Indian farmers?

If a farmer was not aware of the PVP and FR Act at the time of using any protected variety, he will be exempt from the infringement and will be asked not to do such act in future. This right has been provided to Indian farmers by PVP and FR Act 2001.

17. What the purpose of farmers' rights?

Farmers' Rights, as endorsed by FAO in 1989, to recognize that farmers and rural communities have contributed and continue to contribute to creation, conservation, exchange and enhancement of genetic resources and they should be recognized and strengthened in their work.

18. What the significance of farmers' rights?

The important points about nature and scope of Farmers' Rights are briefly presented as follows:

(i) **Ancient Rights:** Farmers' Rights have a deep historic character. They existed since humans created agriculture to serve their needs, and played vital role in conservation of biodiversity. Farmers are the guardians of these genetic resources, which support the evolution of species. Farmers are the inheritors of the skills and knowledge of the generations that have created this biological wealth. For these contributions, it is necessary to recognize Farmers' Rights.

(ii) **Rights Covered:** Farmers' Rights include the right over resources and associated knowledge, united indivisibly, and mean the acceptance of traditional knowledge, respect for cultures and recognition that these are the basis of the creation of knowledge.

(iii) **Property Rights:** The farmers deserve right to control, the right to decide the future of genetic resources, the rights to define the legal framework of property rights of these resources.

(iv) **Collective Nature:** Farmers' Rights are of an eminently collective nature and therefore are protected by sui generic rights.

(v) **Application:** These rights should have a national application, respecting the sovereignty of each country, to establish local laws based on these principles

(vi) **Rights to Means:** It includes territorial rights, right to land, right to water and air and rights to conserve biodiversity and achieve food security.

(vii) **Active Participation:** The farmers deserve right to participation in the definition, elaboration and execution of policies and programmes linked to genetic resources

(viii) **Rights to Technology:** The farmers deserve right to appropriate technology as well as participation in the design and management of research programmes.

(ix) **Rights to Benefit Sharing:** The farmers deserve right to define the control and handling of benefits derived from the use, conservation and management of these resources.

(x) **Rights on Genetic Resources:** The farmers deserve right to use, choose, store and freely exchange of genetic resources.

(xi) **Rights for Sustainable Agriculture:** The farmers deserveright to develop models of sustainable agriculture that protects the biodiversity and to influence the policies that support it.

19. What is the role of farmers related to PGR?

Farmers are both men and women who have domesticated, developed, conserved and made available plant genetic and other natural resources (land; water; vegetation; animal, bird and fish life) to which they have access in order to obtain a livelihood and ensure the well-being of their family through the provision of basic requirements such as food, fuel and water and income from the land.

Farmers have played key role in conservation of Plant Genetic Resource for thousands of years. The role of farmers in relation to PGR is briefly presented below:

(i) Farmers both men and women have conserved plant genetic resources over the past thousands of years.

(ii) Farmers are rightly called as on site managers of plant genetic resources.

(iii) Farmers have in depth knowledge of the land races and wild relatives of cultivated plant species.

(iv) All modern plant varieties contain only those genes that have originated from farmers' traditional varieties (landraces) or wild crop relatives.

(v) The women and men have historically provided the germplasm upon which scientific plant breeding is based.

(vi) Generally, germplasm is provided by small-scale subsistence farmers to plant breeders whose varieties are then adapted/developed for medium to large scale commercial farmers (Spillane, 1996).

20. What is the role of women related to PGR?

(i) Women's special knowledge of the value and diverse uses of plants for nutrition, health and income has important implications for the conservation of plant genetic resources.

(ii) The women have played significant role not only in terms of labour and skills, but also their decision-making about how natural resources are used to satisfy the multiple needs of rural households.

(iii) In much of the developing world, the on-farm and in-situ conservation and use of plant genetic resources begins with women. As farmers, they are responsible for growing crop plants and collecting seed for next year sowing.

(iv) It is estimated that 90 per cent of the planting materials used in developing countries is produced by the farmers themselves, *i.e.* from saved seed. The selection of wild genetic resources for home planting, of seeds for conserving for next year's planting and of asexually propagated material is usually carried out by women. This selection is a sophisticated process that takes into account many different criteria such as taste, color, palatability and texture; resistance to pests and diseases; adaptation to soil and agro-climatic conditions and so on. The selection criteria are built up over years of experience.

21. What are the similarities between Plant Breeders' rights and farmers' Rights?

Plant Breeders' Rights and Farmers' Rights have some similarities which are presented in Table 17.1

TABLE 17.1: Similarities between Plant Breeders' Rights and Farmers Rights

Sl.No.	*Particulars*	*Plant Breeders' Rights*	*Farmers' Rights*
1.	Registration	Essential	Essential
2.	Duration		
	Trees and vines	18 years	18 years
	Extant varieties etc	15 years	15 years
3.	Enforcement	After Registration	After Registration
4.	Requirements	Novelty, Distinctiveness, uniformity and stability	Same as in PBR
5.	Validity	In country of registration	In country of registration

22. What are the differences between plant breeders' rights and farmers' rights?

Plant Breeders' Rights and Farmers' Rights have some differences which are presented in Table 17.2.

TABLE 17.2: Differences between Plant Breeders' Rights and Farmers Rights

Sl.No.	*Particulars*	*Plant Breeders' Rights*	*Farmers' Rights*
1.	Registration Fee	Charged	Not charged
2.	Annual renewal fee	Charged	Not charged
3.	Penalty for infringement	Charged	Not charged
4.	Community Rights	Not covered	Covered
5.	Authorization	Easy	Not permitted

23. Discuss the future outlook related to PGR.

Plant Genetic Resources are the base materials for modern plant breeding programmes Hence their proper conservation and sustainable use are essential. In future, the following points should be given due importance for proper conservation and utilization of plant genetic resources.

(i) **Assistance to Farmers:** Farmers in all regions of the world should be assisted in their conservation of natural resources and more specifically the selection, conservation, improvement and sustainable use of PGR. Non Governmental Organizations should assist in the entire process as they are close to the communities and have the capacities to provide long-term support. The establishment of strong village level institutions with adequate representation of both women and men farmers can be key in promoting successful programmes.

(ii) **Incentives:** Appropriate incentives should be developed, with the active participation of those concerned, to convince communities, industry and governments of the benefits of conservation and development efforts and to provide financial resources, legal support, and technical capability to develop agricultural systems that are suitable for local agro ecological conditions and, where possible, based on locally available resources. Incentives should be used to enable farmers to divert land, labour and capital towards conserving biological resources and to facilitate the participation of certain groups in activities that promote the sustainable use of those resources.

(iii) **Participation of Women Farmers:** The participation of women farmers and the recognition of their contributions in the design and elaboration of such policies and programmes at all levels is essential. Women farmers should be included in decision making processes at community and regional levels. Farmers, communities and countries, and men and women at all levels should participate in the design and implementation of programmes policies and agreements related to PGR.

(iv) **Self Decision:** Those farmers who develop breeding materials should be empowered to decide, at their discretion, if the materials will be made available to other (*i.e.* external) users.

(v) **Equal Rights to Women:** There should be equal rights for conservation and participation in PGR related programmes and policies, which is often lacking in developing countries. There should not be discrimination of gender. In the past, women farmers could get little benefit from support in terms of information and technology, inputs and credit, extension and research.

(vi) **Participatory Plant Breeding:** Participatory research and plant breeding should be promoted with different socio-economic groups in rural communities in order to improve the productivity of local varieties and to promote marketing of special products derived from these materials.

(vii) **Use of Land Races:** Modern plant breeding has concentrated on developing a small number of varieties that are widely adapted to many environments. Land races should also be used for developing varieties suitable for diverse ecological conditions.

(viii) **International Funds:** There is need for the creation of an Intentional Fund to support in-situ conservation and to benefit rural men and women who produce and conserve PGR. In general farmers are interested in conserving resources but lack of financial support acts as a barrier in their way.

(ix) **Integrated Programs:** PGR conservation should be integrated into other community development programmes It will promote and employment in activities related to PGR building up of local institutions, capacity building) and collecting and dissemination;, information on the status and trends of PGR.

(x) **Benefit Sharing:** There should be equitable sharing of benefits arising from the utilization of PGR, knowledge, innovations and practices.

24. What are the advantages of farmers' rights?

There are several advantages of farmers' rights which are listed below:

(i) Provide recognition to farmers, communities or village for their role in conserving Plant Genetic Resources.

(ii) Provide respect to farmers.

(iii) Provide opportunities to farmers to participate in the policies, projects and programs related to plant Genetic Resources [PGR].

(iv) Lead to benefit sharing when farmers material is used in developing new variety by breeders.

(v) Farmers permission is required to use their plant material.

(vi) Provide rewards to those farmers who are actively engaged in conserving Plant Genetic Resources.

25. What are the limitations of farmers' rights?

There are some limitations of farmers' rights which are listed below:

(i) Mostly farmers' rights are of collective nature.

(ii) The benefit sharing is difficult.

(iii) Farmers do not get any financial support from government for the conservation of Plant Genetic Resources.

(iv) Farmers' Rights are of territorial nature.

Chapter 18

Moral Rights

1. What are moral rights?

Moral rights are rights of creators of copyrighted works generally recognized in civil law jurisdictions and first recognized in France and Germany, before they were included in the Berne Convention for the Protection of Literary and Artistic Works in 1989. Moral rights are a special extension to copyright that gives the originator of a copyrighted work rights over its use. The originator can specify moral conditions on the use or exploitation of their creation, even when the rights are licensed or sold to others.

The term "moral rights" is a translation of the French term "droit moral," and refers not to "morals" as advocated by the religious right, but rather to the ability of authors to control the eventual fate of their works. An author is said to have the "moral right" to control his or her work. The concept of moral rights thus relies on the connection between an author and his/her creation. Authors may seek moral rights protection from State moral rights laws.

2. Explain main points related to moral rights.

Moral rights protection varies around the world. For example, in the United States, moral rights protection is at a lower level that in many European Union countries. France, by way of comparison, has one of the most protective moral rights regimes in the world. The main points related to moral rights are presented below:

(i) These are rights provided to authors and directors of copyrights to protect their personal interest in relation to their creations. These rights are provided beyond those strictly recognized in copyright law.

(ii) Moral Rights were granted to authors under the Berne Convention (1989).

(iii) The creator of the work has the right of attribution and integrity, even after sale or transfer of the copyright. In other words, the creator or author shall have the right to claim authorship of the work even after sale or transfer of the copyright.

(iv) **Transfer:** The copyrights can be transferred or assigned to others, but moral Right cannot be transferred or assigned.

(v) The author has right to object to any distortion, mutilation or other modification which would be prejudicial to his honour or reputation

(vi) Moral rights include the right of attribution, the right to have a work published anonymously or pseudonymously, and the right to the integrity of the work.

(vii) Moral rights protect the personal reputation of the author/creator of copyright work.

(viii) Protection of moral rights ends with the death of the author.

3. Is registration required for moral rights?

Like copyright registration is not compulsory for moral rights. But it is desirable to register moral rights to have legal control over the unauthorised use of moral rights.

4. What are the matters covered moral rights?

All those matters which are covered under copyrights are also covered by moral rights.

5. What is the duration of protection of moral rights?

The term of moral rights is the same as the term of copyright protection. So when copyright protection ends and a work falls into the public domain the moral rights protection ceases at the same time. Moral rights are not transferrable, and end only with the life of the author, whereas copyrights continue up to 70 years after the death of the author.

6. What are the rights covered under moral rights?

Moral rights are separate rights to that of copyright and belong to the author of a work. The following four categories of moral rights are associated with authorship, and are quite distinct from copyright [Copyright Designs and Patents Act 1988].

(i) **Right to Paternity**: This right requires acknowledgment of the identity of the author. In other words, this right states that the authorship should be properly acknowledged.

(ii) **Right to Integrity**: This right prevents derogatory treatment of authors' work. The right of integrity protects an author's work from intentional distortion, mutilation, or other modification that is likely to harm the author's honor and reputation, and the destruction of any work of recognized stature This right also protects authors from having their works associated with products, services, causes or institutions that would harm their honor or reputation

(iii) **Right to Attribution**: This right is not to have a work falsely attributed to an author. The right of attribution allows an author to prevent misattribution of his work. This right prevents the use of one's name on any work which the author did not create. This right expires 20 years after the death of the author.

(iv) **Right to Disclosure or Privacy**: This is the right to withhold publication of an image or broadcast. This requires that the authorship of the work should not be disclosed (*i.e.* remains anonymous) where it may tarnish his image. It is applicable in respect of certain photographs and films only.

The author has right to prevent others to alter his work without his permission, to dictate in what way the work can be displayed and to receive resale royalties. The author can also prevent revision, alteration, or distortion of his work, regardless of who owns the work.

7. What are the moral rights protected in Canada?

In Canada, copyright law provides the creators of works protected by copyright with several "moral" rights. These include:

(i) **The right of association**: creators have legal recourse against unauthorized use of their work in association with products, services, causes or institutions in ways which are prejudicial to their honour or reputation;

(ii) The right to remain anonymous; **[Right of Non-disclosure]**

(iii) The right to use a pseudonym; **[Right of Non-disclosure]**

(iv) The right to be identified with the work as its author; **[Right of Paternity]**

(v) **The right of integrity**: creators are protected against "distortions, mutilations or other modifications" to a work which are prejudicial to their honour or reputation. The law assumes that any changes to a unique work (*e.g.* a painting or sculpture) are prejudicial.

8. What are the moral rights protected in United Kingdom?

In **United Kingdom**, the Copyright, Designs and Patents Act 1988 came into force on August 1, 1989. The following five categories of moral rights are protected.

(i) **Identification as author:** This gives creators the right to be identified as author, whenever the work is published commercially, exhibited in public, or included in a TV programme or film. The artists whose works are exhibited in public, especially sculptors and artists/craftspeople whose works are permanently sited in or on buildings and in spaces open to the public, can insist on their identification as author.

(ii) **No unjustified modification:** This right prevents unjustified modification of author's work whenever the work is published commercially, exhibited in public, or included in a TV programme or film. Modification means any addition, deletion, alteration or adaptation of the art work which prejudice the honour and reputation of the author.

(iii) **No false attribution as original author:** This right gives creators the right not to be named as authors of works which they did not create, whenever such works are shown to the public, through exhibition or publication of copies including film and TV programmes.

(iv) **No false attribution as author of altered work:** This right gives creatots the right not to be named as author in the work that has been without his consent. This also prevents anyone who deals with such work or copies of such work by way of trade.

(v) **No false attribution as author of a copy of work:** This right enables artists to prevent dealers falsely trading in copies of their works purporting to have been authorised by them.

9. What is the duration of protection of moral rights in United Kingdom?

All five rights will last for the author's lifetime and for a period after death. The first two, the right to be identified and not to suffer unjustified modification, will continue for 50 years beyond the author's death and will be capable of being left in artists' wills. The last three rights concerning false attributions, will continue for 20 years beyond the author's death but will not be capable of being left in an artist's will exceptionally, in the case of photographs, all five rights will only last for 50 years from the making of the image.

10. Is it possible to transfer moral rights?

None of the moral rights will be capable of being transferred or assigned by the artist to anyone else during his/her lifetime.

11. What are the requirements for protection of moral rights?

Since moral rights are linked with copyrights, there are no special requirements for protection of moral rights.

12. Is it possible to authorize or transfer moral rights?

Moral rights related to copyright work are considered personal to the author and that cannot therefore be bought, sold or transferred. The moral rights cannot be assigned, but can be waived contractually. Under Canadian copyright law a creator cannot assign moral rights. It is only possible for a creator to agree not to exercise moral rights by "waiving" the rights. Authors may, however, waive their moral rights if do so in writing. Moral rights can be waived for one particular individual or for one particular use.

13. When does the moral rights become effective?

The moral rights become effective immediately after release of the copyrighted works.

14. How infringement of moral rights is determined?

The determination of infringement involves two types of components, *viz.* a subjective and an objective component. The subjective component is whether the author himself feels that his work is used in a manner which is prejudicial to his honour or reputation. The object component is an evaluation of public or expert opinion. The word of the author alone is not enough to establish infringement. To sustain a claim of infringement, an author must be supported by expert or public opinion. There are various mechanisms with which an author may enforce his moral rights. These include the following:

(i) An author may show that, in altering or distorting her work, someone has created a "derivative work," thereby violating the Copyright Act.

(ii) If authorship of a work is attributed to an author against her will, or misattributed, the author may have a state action for defamation against the person responsible for the attribution.

(iii) If a person uses the identity of an author, or the works of the author, for her own benefit without the author's permission, then she may have violated the author's right of publicity or may be guilty of misappropriation of the author's work.

The person entitled to any of the moral rights can take legal proceedings in the civil courts for preventing infringements. He can ask for financial compensation for any damage caused by infringer including cost of legal proceedings.

15. What are the similarities between moral rights and mask work rights?

There are some similarities between moral rights and mask work rights which are presented in Table 18.1.

TABLE 18.1: Similarities between Moral Rights and Mask Work Rights

Sl.No.	*Particulars*	*Moral Rights*	*Mask Work*
1	Governed by	*Sui-generis* rights	*Sui-generis* rights
2	Registration	Not compulsory	Not compulsory
3	Validity	Country of registration	Country of registration
4	Infringement	Possible	Possible

16. What are the differences between moral rights and mask work rights?

There are some differences between moral rights and mask work rights which are presented in Table 18.2.

TABLE 18.2: Differences between Moral Rights and Mask Work Rights

Sl.No.	*Particulars*	*Moral Rights*	*Mask Work*
1	Period of protection	Till the life of the author	Unregistered 2 years Registered 10 years
2	Protection over	Copyrighted matters	Integrated Circuits
3	Enforcement	After release of copyrighted work	After creation
4.	Requirements	Copy righted material	Originality of mask work
5	Transfer	Not permitted	Permitted

17. What are the advantages and disadvantages of moral rights?

There are some advantages and disadvantages of moral rights which are listed below:

Advantages

(i) Registration is not compulsory.

(ii) Protection is life long.

(iii) It promotes creative work.

(iv) It provides benefits to the owner.

Disadvantages

(i) It is not transferable.

(ii) It does not provide legal rights without registration.

(iii) It may lead to monopoly.

Chapter 19

Supplementary Protection Certificates

1. What is supplementary protection certificate?

A supplementary protection certificate (SPC) is a special form of intellectual property (IP) that extends the protection of a patented active ingredient or combination of active ingredients present in a pharmaceutical or plant protection product after the expiry of the patent. In other words, a supplementary protection certificate (SPC) is a *sui-generis* form of intellectual property right [legal protection] that extends the protection conferred by a patent [called basic patent] in respect of medicinal products, such as drugs, and plant protection products, such as insecticides, and herbicides.

2. What is the need for supplementary protection certificate?

A pharmaceutical or plant protection product must obtain: (i) regulatory approval, (ii) a marketing authorization, and (iii) or product license before its sale in the market. The grant of such an official authorization does often take years because of lengthy testing procedure. Due to the long term procedure for an authorization a patent proprietor may lose valuable time, in which he would have been able to market his product. The SPC compensates for the delay in authorization.

3. Explain briefly main points related to supplementary protection certificate.

It is estimated that more than 8,000 SPCs for medicinal and plant protection products have been filed in Europe between 1991 and 2003. Main points related to supplementary protection certificate are given below:

(i) It is in use in European Union member countries.

(ii) It is a *sui-generis*, patent like, intellectual property right.

(iii) The SPC comes into force only after the expiry of corresponding patent.

(iv) **Duration:** It has maximum life time of 5 years when granted after 10 years after filing date of a patent. The term would be less than 5 years if it is granted after 5 years from the date of filing the corresponding patent.

(v) **Conditions of Issue:** It is issued against a patent generally immediately on expiry of the term of the patent. It cannot be issued before 5 years from filing date of corresponding patent

(vi) A separate application is to be filed for registration in each country.

(vii) It extends the protection conferred by the basic patent beyond the term of that patent but in respect of medicinal or plant protection products covered by authorization. It does not extend the term of the patent itself.

A separate certificate is issued for each product, even if they are protected by the same basic patent.

4. What is the procedure for registration of supplementary protection certificate?

The registration is required which is issued on national basis. The application is to be submitted in the prescribed application form in the patent office. If the application does not meet the prescribed conditions, the applicant is asked to eliminate the deficiencies within an appropriate term (about two months). If the deficiencies are not eliminated within the term set, the application shall be rejected. Generally, the above mentioned term is extendable. Thus the registration process consists of following steps:

(i) **Filing of application**: The application complete in all respects is submitted in the patent office along with prescribed fee.

(ii) **Examination of application**: The application is examined to check whether it meets requirement for grant of supplementary protection certificate.The application is examined to ensure the following things:

- (a) The application is accompanied by the prescribed fee,
- (b) The application has been filed within the prescribed period,
- (c) The application contains the required particulars and documents, and
- (d) The basic patent is in force.

(iii) **Issue of Certificate:** If the application meets the desired requirement, the SPC is issued for a specific period for which annual maintenance fee is to be paid every year.

5. What are the matters covered supplementary protection certificate?

In European Union member countries, a **supplementary protection certificate** (short certificate) is a *sui-generis*, patent-like, intellectual property right. This type of right is available for medicinal products such as drugs and plant protection products, such as insecticides, fungicides and herbicides. The certificate is granted for the products which constitute (i) the active ingradient or combination of active ingradients of a medicinal product or (ii) the active substance or combination of active substances of a plant protection product.

6. What is the duration of protection of supplementary protection certificate?

The maximum life-span of the Supplementary Protection Certificate is five years from its effective date. The term of the certificate may not exceed five years. The term of an SPC depends on the date of issuance of the first market authorisation.

(i) If the first market authorisation is issued less than five years after the filing date of the corresponding patent, no SPC is granted.

(ii) If the first market authorisation is issued more than five years but less than ten years after the filing date of the corresponding patent, an SPC is granted for a term corresponding to the period elapsed between the five-year point and the market authorisation issuance date.

(iii) If the first market authorisation is issued more than ten years after the filing date of the corresponding patent, an SPC is granted for a five-year term.

The application for the certificate has to be filed within six months of the date on which the patent is expiring. The supplementary Protection Certificate extends the life-span of a basic patent up to 5 years.

7. What are the rights covered by supplementary protection certificate?

A certificate may only be granted to the holder of the basic patent or his successor in title. The certificate confers the same rights as conferred by the basic patent and is subject to the same limitations and the same obligations.

8. What are the requirements for registration of supplementary protection certificate?

The following conditions should be fulfilled for grant of a supplementary protection certificates:

(i) The basic patent protecting the product should in force,

(ii) The product has not already been the subject of a certificate,

(iii) A valid authorization has been granted to place the product on the market.

(iv) The authorization should be the first authorization to place the product on the market.

9. What are the fees payable for registration of supplementary protection certificate?

The prescribed processing fee is deposited along with the application for a supplementary Protection Certificate. An annual fee is payable for each year of the effective period of the certificate. Any period of less than 12 months is considered as a whole year. For instance, an effective period of 3 years and 6 months will require the payment of 4 years' annual fees. Annual fees are paid for maintaining a supplementary protection certificate. The yearly fees are paid every year three months before the due date and up to six months after the due date. If the yearly fees are paid after the due date, then an additional fee of 20 per cent is due with the yearly fee. The annual fees for a certificate are to be paid from the first year.

10. Is it possible to transfer a supplementary protection certificate?

The owner of the supplementary protection certificate has the power to transfer the right of certificate to any other person if he so desires. In other words, the owner or inventor may transfer ownership to any one. Employee often assign such rights to their employers as part of their employment contracts.

11. When does a supplementary protection certificate become effective?

A supplementary protection certificate comes into force only after the expiry of a basic patent. In other words, it enters into force when the related patent expires. The followings are also required before enforcement.

(i) The basic patent has not previously lapsed or been revoked, and
(ii) The annual fees are paid in time.

12. What is the validity of a supplementary protection certificate?

A Supplementary Protection Certificate does not give protection abroad; it is only effective in the country of registration. Applications for a supplementary protection certificate must be filed on a country-by-country basis. There is no unitary European supplementary protection certificate, but national ones only. The certificate will lapse under following situations:

(i) At the end of the effective period,
(ii) If the holder surrenders it,
(iii) If the annual fees have not been paid in time, and
(iv) If the product covered is no longer placed on the market.

A lapsed certificate may be restored when a new authorization to place the product on the market is granted. If a certificate is declared invalid, the fee for the remaining period will be refunded on written request by the holder.

13. How the infringement of a supplementary protection certificate can be prevented ?

Like that of a basic patent, supplementary protection certificate provides legal rights against infringement of the protected materials.

Documents provided in support of an application for a certificate including marketing authorizations, will normally be made open to public inspection after they are filed or sent to the office.Copies of any documents which are not treated as confidential will be available upon request as in case of documents relating to patents.

14. What are the differences between SPC and moral rights?

There are several differences between supplementary protection certificate and moral rights which are presented in Table 19.1.

TABLE 19.1: Differences between Supplementary Protection Certificate and Moral Rights

Sl.No.	*Particulars*	*Supplementary Protection Certificate*	*Moral Rights*
1	Registration	Compulsory	Not compulsory
2	Period of protection	Five years	Till the life of the author
3	Matters covered	Medicinal products and plant protection products	Copyrighted matters
4	Validity	Country of registration	Country of registration
5	Enforcement	After registration	After release of copyrighted work
6	Requirements	Patented matters	Copy righted material
7	Transfer	Permitted	Not permitted

15. What are the advantages and disadvantages of SPC?

There are some advantages and disadvantages of supplementary protection certificate which are listed as follows:

Advantages

(i) Registration is not compulsory.
(ii) It extends life span of a patent by 5 years.
(iii) It promotes research work on pharmaceuticals and pesticides.
(iv) It provides benefits to the owner.

Disadvantages

(i) It has short life span [5 years].
(ii) It does not provide legal rights without registration.
(iii) It may lead to monopoly.

Chapter 20

Indigenous Property Rights

1. What is traditional knowledge?

The Traditional Knowledge [TK] refers to the knowledge, innovations and practices of indigenous people and local communities. In other words, it is the: knowledge gained through tradition from ancestors about plants and animals, medicines, agricultural practices and other matters. Traditional knowledge is also known as indigenous knowledge or local knowledge or traditional ecological knowledge.

2. Describe the main points related to traditional knowledge.

Main points about traditional knowledge are listed below:

(i) Traditional knowledge can be protected by *sui-generis* rights

(ii) It covers the matured long-standing traditions and practices of certain regional, indigenous, or local communities. In other words, it is gained from experience gained over the centuries and adapted to local culture and environment.

(iii) It includes knowledge about plants, animals, indigenous treatment of diseases through herbal medicines, cultural practices,*etc.*

(iv) Traditional knowledge is based on teachings and experiences and is orally passed on from generation to generation to some persons only. Earlier there were no written document on traditional knowledge.

(v) Some forms of traditional knowledge are expressed through stories, legends, folklore, rituals, songs and even laws.

(vi) There is difference between local knowledge and traditional knowledge. The former existed from decade to century, whereas the latter existed from millennia.

(vii) It is collectively owned by the people or natives of an ecosystem or region.

(viii) Traditional knowledge incorporates knowledge of ecosystem and code of ethics governing appropriate use of environment. This code includes rules and conventions promoting desirable ecosystem relations, human-animal interactions and even social relationships. A combination of traditional and non-traditional knowledge will form a rich and distinctive understanding of life and the world.

(ix) The ecological knowledge is composed of both traditional knowledge and experiential knowledge [gained through personal experience].

Traditional knowledge is very important almost in every sphere of life. Hence, its protection is very necessary otherwise this valuable knowledge will be gradually lost

3. What are different types of traditional knowledge?

The traditional knowledge has been classified into six major fields such as traditional knowledge of (i) agriculture, (ii) horticulture, (iii) forestry, (iv) medicines, (v) fisheries and human culture. These are briefly discussed as follows:

(i) **Agriculture:** Traditional knowledge related to agriculture includes following aspects.

(a) It includes knowledge of plant and animal genetic resources and their conservation.

(b) Knowledge of production practices including development of plant varieties and animal breeds.

(ii) **Horticulture:** Traditional knowledge related to horticulture includes following aspects.

(a) Knowledge of horticultural plant genetic resources and their conservation.

(b) Knowledge of production and improvement of horticultural plant species.

(iii) **Forestry:** It includes knowledge of timber yielding and other useful forest plants and their conservation.

(iv) **Medicines:** It include knowledge of medicinal plant species which are used for treating human and animal diseases.

(v) **Fisheries:** It includes in depth knowledge about fish genetic resources and their conservation.

(vi) **Human Culture:** It includes traditional knowledge which is expressed through stories, legends, folklore, rituals, songs and even laws.

Moreover, local people have better knowledge of environment such as snow, ice, weather, resources and the relationship between things. In other words, local people have better insight into the ecosystem of their region or locality.

4. Is registration of traditional knowledge compulsory?

Registration of traditional knowledge is not essential. However, registration will provide legal rights to the people of local communities to protect their traditional knowledge.

5. What is the duration for protection of traditional knowledge?

The traditional knowledge can be protected for indefinite period by the local communities. In other words, there is no limit for conservation of traditional knowledge.

6. What is the validity of traditional knowledge?

The traditional knowledge is general used in the area of its conservation by local people.

7. What are the areas covered by traditional knowledge?

The traditional knowledge is mainly of practical nature, particularly in such fields as agriculture, horticulture, forestry, fisheries, animals and herbal medicines. It also includes cultural matters such as stories, songs, folklore, proverbs, beliefs, cultural values, rituals, community laws and local language, *etc.*

8. How the traditional knowledge is transmitted?

The knowledge is directly transmitted from individual to individual. Generally, it is passed on from father to son and mother to daughter generation after generation. Local people or natives possess both collective and individual traditional knowledge. The traditional knowledge is passed on by elders to the young. As the elders die, the some knowledge which has not been passed on is lost.

9. Who has the control over the traditional knowledge?

The native people have the rights to prevent use of their traditional knowledge by others without permission.

10. Who own the rights of traditional knowledge?

The local people or natives, who have conserved the knowledge for centuries together, own the intellectual property rights to their traditional knowledge, even if much of it is yet to be written down. No one has the right to document or use traditional knowledge without permission. In other words, the following things are required for use of traditional knowledge:

(i) Prior permission of natives.

(ii) The source of traditional knowledge should be acknowledged

(iii) It should be properly interpreted.

(iv) There should respect for traditional knowledge.

11. What is bio-piracy?

The illegal collection of indigenous plants by corporations who patent them for their own use is called bio-piracy. There is bio-piracy of traditional knowledge which needs to be prevented.

12. How the benefits of traditional knowledge can be shared?

Local people possess both collective and individual traditional knowledge. Most traditional knowledge is shared among community members. But some traditional knowledge may be specific to an individual. This type of traditional knowledge is kept secret by some people who are engaged in treating some human and animal diseases. Moreover, it depends on the individual to decide whether he wishes to share his knowledge with outside parties or not.

13. What is the approach of traditional knowledge?

Traditional knowledge assumes a holistic view including language, culture, practices, spirituality, mythology, customs and even the social organization of local people. It can neither be divided nor separated from the local people who hold it. It is rooted in the spiritual health, culture and language of people. It is a way of life. In other words, it cannot be separated into secular[non-religious] and sacred knowledge.

14. What is the system of traditional knowledge?

Traditional knowledge is an authority system. It sets out the rules governing the use of resources, respect and an obligation to share. It is a dynamic, cumulative and stable system. Native people consider it truth.

15. Who are adopting traditional knowledge ?

The traditional knowledge is adopted mainly in local areas where it has been developed. On the other hand, scientific knowledge is used in a larger area. In other words, traditional knowledge has limited practical applications as compared to scientific knowledge.

16. What are the basis for explanation of traditional knowledge ?

The explanation of traditional knowledge is based on examples, stories and myths, whereas the explanation of scientific knowledge is based on hypothesis, theories and law.

17. Describe value of T.K. and role of local people.

There is increasing appreciation of the value of traditional knowledge. This knowledge is valuable to the indigenous as well as to the modern industry and agriculture. The role of traditional knowledge is briefly presented below:

(i) **Medicines:** Many widely used products such as plant based medicines and cosmetics are derived from traditional knowledge.

(ii) **Handicrafts:** Handicraft and non-wood forest products are based on TK.

(iii) **Sustainable use:** Most indigenous and local communities live in areas which very rich in plant genetic resources. However, local people have made sustainable use of such resources for thousands of years. They have not exploited the resources.

(iv) **Conservation of biodiversity:** The local people are rightly called as natural resource managers. They are managing plant and animal genetic resources for thousands of years. Women play a key role in the conservation of plant genetic resources as they are the primary harvesters of medicinal plants, seed stocks and small grains.

(v) **Special skills and Techniques:** The indigenous people, as on site communities, have special skills and extensive knowledge of local environments. Their knowledge provides valuable information to the global community and useful model for biodiversity policies.

18. What are the differences between traditional knowledge and scientific knowledge?

There are several differences between traditional knowledge and scientific knowledge. A brief comparison of both these is presented in Table 20.1.

TABLE 20.1: Comparison of Traditional knowledge and Scientific Knowledge

Sl.No.	*Particulars*	*Traditional Knowledge*	*Scientific Knowledge*
1	Assumption as	The truth	Best approximation
2	Type	Sacred and interconnected	Secular and segregated
3	Teaching through	Story telling	Formal education
4	Learning by	Work and experience	Formal education
5	Format	Oral and visual	Written
6	Approach	Holistic	Analytical [simpler]
7	Based on	Intuition and experience	Model or theory or facts
8	Nature	Subjective	Objective
9	Transfer	Time taking	Fast
10	Explanation based on	Examples, stories and myths	Hypothesis, theories and law
11	Credibility	Collective to local people	To individual inventor
12	Adoption	In local areas	In entire country

19. How would you protect traditional knowledge?

The role of intellectual property (IP) systems in relation to traditional knowledge (TK), and how to preserve, protect and equitably make use of TK, has recently received increasing attention at global level. The major issues are food and agriculture, the environment, notably the conservation of biological diversity, health, including traditional medicines, human rights and other Indigenous matters. Thus protection of traditional knowledge has become very important. The protection of traditional knowledge can be of two types, *viz.* (i) defensive protection, and (ii) positive protection. These are explained as follows:

1. **Defensive protection:** This includes those measures which ensure that intellectual property rights over traditional knowledge are not given to parties other than the customary traditional knowledge holders. Some countries and communities are developing TK databases that may be used as evidence of prior art to defeat a claim to a patent on such TK.
2. **Positive protection:** This includes the creation of positive rights in traditional knowledge that empower traditional knowledge holders to protect and promote their traditional knowledge. In some countries, *sui-generis* legislation has been developed specifically to address the positive protection of traditional knowledge. Traditional knowledge providers and users may enter into contractual agreements and/or use existing Intellectual Property systems of protection.

For success of any project, program or policy on traditional knowledge, the following aspects should be given due importance.

(i) There should be effective participation of indigenous and local communities in decision making and policy planning.

(ii) The traditional knowledge should be given due respect and importance, valued, preserved and properly maintained such as the knowledge relevant to conservation and sustainable use of biodiversity.

(iii) The wider application of traditional knowledge should be made with the approval and involvement of the concerned indigenous and local communities.

(iv) There should be equitable sharing of the benefits arising from the utilization of traditional knowledge.

Many natives do not want to share their traditional knowledge in depth with outsiders because they fear that their traditional knowledge will be extracted, misappropriated and misused without respect to them.

20. What are advantages and disadvantages of traditional knowledge?

There are some advantages and some disadvantages of traditional knowledge which are listed below:

Advantages

(i) There is no need of registration.

(ii) Native people have the right to their traditional knowledge.

(iii) It will help natives in profit sharing for benefits by users.

(iv) It is useful in sustainable use of resources.

(v) It brings respect and reputation to the local people

(vi) It is holistic approach to knowledge.

Disadvantages

(i) Non sharing of TK will promote monopoly.

(ii) It is some times lost due to non-disclosure.
(iii) It is passed on orally to next generation.
(iv) It is based on personal experience.
(v) Explanations are based on examples, stories and myth.

Appendices

Appendix 1: Summary of Intellectual Property Rights

Sl.No.	*Type of IPR*	*Registration*	*Period Years*	*Authori-zation*	*Requirements*
A	**Primary Rights**				
1	Copyrights	Not essential	70	Yes	Originality Fixed expression
2	Patents	Essential	20	Yes	Novelty Usefulness Non-obviousness
3	Trademarks	Not essential	10	Yes	Distinctiveness No multiple appearances
4	Industrial Design	Essential	05	Yes	Novelty Commercial use
5	Utility Models	Essential	6-10	Yes	Novelty Industrial application
6	Geographical Indications	Not essential	10	No	Special quality Reputation
7	Trade Secret	Not essential	Indefinite	Yes	Confidential Commercial use
8	Related Rights Performers and Database creators	Not essential	15	Yes	Originality Usefulness
	Phonogram producers and Broadcasting	Not essential	20	Yes	Originality Usefulness

	Photographers and designers	Not essential	25	Yes	Originality Usefulness
	Film Producers	Not essential	50	Yes	Originality Usefulness
9	Domain Names	Essential	01	Yes	Unique Internet Web address
10	Trade Names	Not essential	03	Yes	Novelty Distinctiveness
B	***Sui-generis* Rights**				
11	Database Rights	Not essential	15	Yes	New creation and Investment
Sl.No.	*Type of IPR*	*Registration*	*Period Years*	*Authori-zation*	*Requirements*
12	Mask Work Rights Registered Unregistered	Not essential	 10 02	Yes	Originality and Usefulness
13	Plant Breeders' Rights Annual Plants Perennial Plants	Essential	 15 18	Yes	Novelty Distinctiveness Uniformity Stability
14	Moral Rights	Not essential	Lifelong	Yes	Linked with copyrights
15	SPC	Essential	05	Yes	After expiry of basic patent
16	Traditional Knowledge	Not essential	Indefinite	No	Large scale usefulness

Appendix 2: Address of Patent, Copyright and Trademark Offices in India.

Sl.No.	Name of IPR	Head Office/Branch Office	Location
1	Patent	Head Office	Nizam Palace, Kolkata.
		Branch Unit	Karol Baugh, New Delhi.
		Branch Office	Parel West, Mumbai.
		Branch Office	Wallajah Road, Chennai.
		Patent Information System	CGO Complex, Seminary Hills, Nagpur.
2	Patents, Designs and Trademarks	Controller General of Patents, Designs and Trademarks	Maharishi Kurvey Road, Mumbai.
3	Copyrights	Registrar of Copyrights	Ministry of HRD, Kasturba Gandhi Marg, New Delhi.
4	Trademarks	Head Office	Old Central Building, M.K. Road, Mumbai.
		Branch Office	Ashram Road, Ahmedabad.
		Branch Office	Chowringhee Square, Kolkata.
		Branch Office	Okhla Industrial Estate, New Delhi.
		Branch Office	Basant Nagar, Chennai.

Appendix 3: Country Code Top Level Domain [ccTLD] Extensions for some selected Countries.

Sl.No.	Country	Domain Name	Sl.No.	Country	Domain Name
1	Austria	.at	26	Myanmar	.mm
2	Australia	.au	27	Nepal	.np
3	Argentina	.ar	28	Netherlands	.nl
4	Bangladesh	.bd	29	New Zealand	.nz
5	Belgium	.be	30	Nigeria	.ng
6	Brazil	.br	31	Norway	.no
7	Cambodia	.kh	32	Pakistan	.pk
8	Canada	.ca	33	Peru	.pe
9	China	.cn	34	Philippines	.ph
10	Colombia	.co	35	Poland	.pl
11	Cuba	.cu	36	Saudi Arabia	.sa
12	Denmark	.dk	37	South Africa	.za
13	Egypt	.eg	38	Spain	.es
14	Finland	.fi	39	Sri Lanka	.lk
15	France	.fr	40	Singapore	.sg
16	Germany	.de	41	Sweden	.sd
17	Greece	.gr	42	Switzerland	.ch
18	Hungry	.hu	43	Thailand	.th
19	India	.in	44	Turkey	.tr
20	Indonesia	.id	45	Uganda	.ug
21	Israel	.il	46	UAE	.ue
22	Italy	.it	47	United Kingdom	.uk
23	Japan	.jp	48	USA	.us
24	Malaysia	.my	49	Vietnam	.vn
25	Mexico	.mx	50	Zambia	.zm

Glossary

Abandonment: An application for trade-mark or copyright registration may be considered abandoned if the applicant does not take steps to complete the process. An application for Industrial design will be considered abandoned if the applicant fails to reply within the specified time to any report that sets out objections to registration.

Abstract: A brief summary of an invention.

Abuse of Patent: Abusing one's patent rights by failing to make one's invention commercially available.

Advertisement: Publication of a trade-mark application in the *Trade-marks Journal*. Details are published to allow opportunity for challenges to the application (opposition).

Allowance: Acknowledgement by the Trade-marks Office that an application is eligible for registration.

Application: The formal request for registration of intellectual property.

Artistic Work: Any visual representation such as a painting, drawing, map, photograph, sculpture, engraving or architectural plan.

Assignment: Transfer of intellectual property rights from the owner to another party.

Author: The creator of an artistic, literary, musical or dramatic work.

Berne Copyright Convention: An international treaty extending copyright protection in member countries to nationals of other member countries. Canada is a signatory to this treaty.

Biodiversity: Variety of biological organisms found in the entire world or it the some of genes, species and ecosystem in a region.

Breeders' Exemption: Rights provided to Plant Breeders use protected material as the basis to develop a new variety or for other research work; also called research exemptions or breeders' privilege.

Cancopy: A reprography collective that grants licenses to photocopy works and collects fees on behalf of its members.

CBD: Convention on Biological Diversity.

Certificate of registration: Official confirmation that your design, copyright or trade-mark has been registered.

Certification Marks: Marks identifying goods or services meeting a defined standard (*e.g.*, Wool Mark on clothing).

Claims: The basis for filing an application that allows for the determination of the person entitled to the registration of the mark.

Clearly Descriptive: A mark that clearly describes a feature of a ware or service and that is therefore not registered as a trade-mark.

Coexistence Agreement: A binding agreement whereby two or more traders agree how to operate in the same or similar market to avoid any potential confusion.

Commercial Cultivar: A variety which is used for cultivation on vast area.

Candidate Variety: A variety which is to be ptotected under Plant Variety Protection Act.

Copyright: A state monopoly granted to protect artistic, musical and literary works.

Collective: Organization that administers rights granted by the copyright system on behalf of copyright owners who have joined that collective.

Compulsory License: The right granted by the Commissioner of Patents to one or more parties to produce a patented invention. These are granted under certain circumstances as a measure against abuse of patent rights.

Copyrights: Rights that Provide protection for literary, artistic, dramatic or musical work, computer programs, and sound recordings.

Copyright Act: The legislation governing copyright in in a country.

Copyright Declaration: Formal statement that the applicant is the author, the owner of the work, an assignee or a person to whom an interest in the copyright has been granted by license.

Copyright Infringement: Violation of copyright through unauthorized copying or use of a work or other subject matter under copyright.

Copyright Office: The central government office responsible for registering copyrights. **Crown copyright:** Copyright of works prepared for or published by the government, *i.e.*, government publications.

Database: A collection of data organized for rapid search and retrieval by a computer.

Deceptively Mis-descriptive: A mark that deceptively mis-describes a feature of a ware or service in such a way as to potentially mislead the consumer is not registered as a trade-mark.

Declaration: Formal statement that you are the proprietor of the design and that, to your knowledge, no one else was using the design when you created it.

Description: A description is a basic requirement of an application for an industrial design. It identifies the features of shape, configuration, pattern or ornament that distinguish the design, and indicates whether the design resides in the entire article or a portion.

Design: The appearance of the whole or part of a product, including any pattern or texture applied to a surface.

Design Patent: The term used in the United States to describe a form of protection similar to that of industrial design protection in Canada.

Disclaimer: A statement that the applicant does not claim exclusive rights for a portion of the trade-mark.

Distinctiveness: The variety must be distinguishable in at least one character from previously available varieties.

Distinguishing: The unique shape of a product or package that can be registered as a trade-mark.

Domain Name: The virtual address on the Internet.

Dramatic Work: Includes plays, screenplays, scripts, films, videos and choreographic works as well as translations of such works.

Due Diligence: Investigations undertaken to assess the strength of one or more intellectual property rights.

Drawings: Drawings (or photographs) disclose the industrial design and are a basic requirement of a design application.

Ecosystem Diversity: Biological variation found in the ecosystems in a country.

EPO (European Patent Office): The pan-European body responsible for granting patents across Europe.

Examination

Trademarks: The process through which the Trade-marks Office determines whether an application for trade-mark is registrable.

Industrial Designs: The process through which the Industrial Design Office determines whether a design warrants registration. The main objective is to determine whether the design is indeed design subject matter, that it is original, and that the documentation meets the requirements of the *Acts* and *Rules*.

Exception: A provision in a copyright law permitting the use of a work by defined user groups without the consent of its creator or without the payment of royalties, conditions that would normally constitute an infringement of copyright. Examples of user group beneficiaries from exceptions are educational institutions, libraries, museums, archives and persons with a perceptual disability.

Example Variety: In DUS, testing, a variety which is used for comparing a particular trait.

Exotic Variety: A foreign variety which is directly recommended for commercial cultivation.

Extant Variety: All released and notified varieties which have not been protected under plant variety protection act. It includes notified varieties, private varieties, farmers' varieties and varieties of common knowledge.

Fair Dealing: Use of works for purposes of private study, research, criticism, review or news reporting that does not constitute infringement of copyright.

Farmers' Exemption: Legal rights provided to farmers to save, use, exchange, share or sell his farm produce of a protected variety; also called farmers' rights or farmers' privilege.

Farmers' Rights: The legal rights provided to farmers to save, use, sow, replant, exchange, share or sell his farm produce including seed of a variety protected under Plant Variety Protection Act.

Farmers Variety: A variety developed by farmers and used for cultivation for several years.

Fees: Specific money payable for various services provided by Intellectual Property Office.

First to file: A patent system in which the first inventor to file a patent application for a specific invention is entitled to the patent. In most of the countries, the first person to file has priority over other people claiming rights for the same invention.

Genetic Diversity: The variety of genes and genotypes found in a particular species.

Globalization: The process of organizing the whole world into a single integrated marketing unit.

Green Revolution: Quantum jump in the productivity of cereals by evolving high yielding varieties.

Improved Cultivars: Varieties developed by scientific method and released for commercial cultivation by central or State variety release committee.

Industrial Design: The visual features of shape, configuration, pattern or ornament applied to a manufactured article.

Industrial Design Act: Legislation governing registration and ownership of industrial designs in a country.

Industrial Design Office: The central government office responsible for registration of industrial designs.

Infringement: The unauthorized copying or use of a protected form of intellectual property. It may include unauthorized use of industrial design, trademarks and patents.

Industrial Design: Violation of industrial design rights through unauthorized use of a design.

(i) **Trade-marks:** Violation of trade-mark rights through unauthorized use of a trade-mark.

(ii) **Patents:** Unauthorized use or sale of a protected invention.

Integrated Circuit Topography: (a) The three-dimensional configuration of the electronic circuits embodied in integrated circuit products or layout designs. (b) Legal protection against imitation of this configuration.

Intellectual Property: A form of creative work that can be protected through a trade-mark, patent, copyright, industrial design or integrated circuit topography.

Intellectual Property Office: The government agency responsible for the administration of intellectual property laws.

Intellectual Property Right: Legal rights granted to protect a form of intellectual property for *e.g.* a patent right, design right, trade mark right or copyright.

Intellectual Property Assets: Those intangible assets that derive from intellectual property and that, typically, add value to your business.

Intellectual Property Auditing: Investigations that are undertaken to determine the nature and number of Intellectual Property Rights owned by a business.

Invention: An innovation that is both novel and non-obvious.

Landraces: The primitive cultivars which were selected and cultivated by farmers for many generations.

Letter of Consent: Letter from one party, typically the owner of a stronger trade mark, consenting to the use and/or registration of another party's trade mark.

Literary work: Work consisting of text such as novels, poems, song lyrics without music, catalogue, reports, tables as well as translations of such works. It also includes computer programs.

License: Legal agreement granting permission to any person to use Intellectual Property Rights in exchange for some consideration or benefit. A license does not change ownership of the copyright.

License: A person other than the owner having the same rights as the owner for use and sale of a product, mark or design.

Licensing Design: Granting of an exclusive license from the owner of the design to another party that allows said party to manufacture, use and sell the design, and to sub-license others to do so.

Licensing invention: Allowing a business or individual to manufacture and sell an invention, usually in exchange for royalties.

Madrid Protocol: An international agreement for the registration of trade marks in a number of countries by means of a single application.

Mask Work: A two or three-dimensional layout or topography of an integrated circuit (IC or "chip").

Maintenance fee:

(i) **Industrial design:** Fee required in order to maintain the rights to an for a second five year period.

(ii) **Patents:** Annual fee to keep a patent.

Marking:

(i) **Copyrights:** Indicating copyright with a small "c" in a circle (©), the name of the copyright owner and the year of first publication.

(ii) **Industrial design:** For registered design a capital "D" is placed in a circle along with the name, or abbreviation thereof, of the design's proprietor on the object to which the design has been applied, or to its label or packaging.

(iii) **Trade-marks:** Trade-mark owners often indicate their registration through certain symbols, such as R in a circle (registered), TM (trade-mark), SM (service mark).

Modern Cultivars: The currently cultivated high yielding varieties.

Monopoly Right: Another term for Intellectual Property Right; the period during which a patent, design, trade mark or copyright can be exclusively exploited by its owner.

Moral Rights: Rights an author retains over the integrity of a work and the right to be named as its author even after sale or transfer of the copyright.

Musical Work: Work that consists of music plus lyrics or music only.

Nom-de-plume: A name used by a writer instead of his or her real name.

Novelty: The newness of a invention, product, marks, variety, *etc.*

OHIM (Office for Harmonization in the Internal Market): The European Community's Trade Mark Registration Authority.

Opposition: The process whereby members of the public may object to the granting of a trade-mark registration if they have valid grounds for doing so.

Patent: A legal right provided to the inventor to exclude others from making, using or selling his invention.

Patent Agent: A fully qualified under the rules of the UK's Chartered Institute of Patent Agents.

Patent application: A technical document describing in detail an innovation for which a patent is sought.

Patent Attorney: A fully qualified to act before the Patent Office in a country.

Patent Cooperation Treaty: An international treaty providing for standardized filing procedures for foreign patents in the countries that have signed the treaty.

Patent Examiner: An official with technical expertise charged with the task of classifying a patent or an application, or of determining whether a patent application fulfills the requirements for a grant of patent.

Patent Office: The patent granting authority and disseminator of patent information.

Patent Office Action: An official written communication by the Patent Office on the merits of an application.

Patent pending: A label sometimes affixed to new products informing others that the inventor has applied for a patent and that legal protection from infringement (including retroactive rights) may be forthcoming.

Performers and Producers Rights: A term used to indicate rights of performers and sound recording producers to be remunerated when their performances and sound recordings are performed publicly or broadcast. Also referred to as "neighbouring rights".

Performing Rights Society: An organization that administers rights to perform musical works on behalf of composers, lyricists, songwriters and music publishers.

Piracy: An unauthorized duplication of goods protected by intellectual property law (*e.g.* copying software unlawfully).

Plagiarism: Using the work (or part of it) of another person and claiming it as your own.

Posthumous Work: A work that is published for the first time (or, for certain types of works, published, performed or delivered in public for the first time) after the author's death.

Popular Variety: A widely grown crop variety.

Preferred Embodiment: An explanation, in a patent application, of the most appropriate and useful practical applications of an invention.

Primitive Cultivars: Same as land races.

Prior art: Industrial Design : The body of registered industrial designs or other published designs including catalogues, manuals and other publications throughout the world relevant to the application for registration of a design.

Private Copying: Copying of pre-recorded musical works, performers' performances and sound recordings into a blank medium, such as audio tape or cassette, for personal use.

Private Copying Levy: An amount of money established by the Copyright Board on blank recording media, whether made here or imported, by manufacturers and importers and is then sold in Canada redistributed to eligible composers, lyricists, performers and sound recording makers through their professional associations and collectives.

Prosecution: All the steps involved in completing a patent application.

Proprietor: The proprietor of an industrial design is the owner of the design. This can be the person who created it or who paid to have it created or who bought the design rights from the original owner.

Publication: Making available copies of a work to the public.

Plant Breeders' Rights: Legal rights granted to the Plant Breeders/owner for the control over the multiplication and sale of their particular plant variety.

Place of Origin: A word or depiction that designates the origin of a product or service and that therefore may not be registered as a trade-mark.

Preliminary Search: The search of Trade-marks Office records that one should undertake before submitting an application for trade-mark registration.

Pre-publication Verification: A second search of Trade-marks Office records to ensure that no confusing trade-marks exist before a given application is published in the *Trade-marks Journal.*

Prohibited Marks: Marks that are specifically prohibited from use in the Trade-marks Act

Proposed Use: A statement indicating intention to use a certain trade-mark and how that use will occur.

Registration: The granting of formal recognition of a trade-mark, copyright, industrial design, or integrated circuit topography by the Trade-marks, Copyright, and Industrial Design Office respectively.

Registered Patent Agent: A specialist entitled to prepare and prosecute patent applications.

Registered Trademark: A trade-mark entered on the federal government's Trade-marks Register, which formally recognizes the owner's rights to the mark.

Registrar of Trademarks: The official responsible for the Trade-marks Office in a country.

Registrar of Copyrights: The official responsible for the Copyright Office in a country.

Registrar of Integrated Circuit Topographies: The official responsible for the Integrated Circuit Topography Office in a country.

Renewal Fee:

(i) **Industrial Designs:** Fee required in order to renew the rights to an industrial design for a second five-year period.

(ii) **Trademarks:** Fee required in order to renew a trade-mark registration for an additional fifteen-year period.

Representative for Service: A person appointed by the applicant or registered proprietor to receive documents on his/her behalf.

Rome Convention Country: A country that is party to the international convention for the protection of performers, producers of phonograms and broadcasting organizations done at Rome on October 26, 1961.

Royalty: A sum paid to copyright owners for the sale or use of their works, or other subject matter.

Search: The act of searching through intellectual property records in order to verify whether a patent, trade-mark or industrial design has been previously filed or registered.

SOCAN: The Society of Composers, Authors and Music Publishers of Canada. SOCAN is the only performing rights society in Canada.

Sound recording: A device which produces sounds, such as cassette, record, or CD player.

Special Order Examination: A request to the Patent Office for early consideration of a patent application.

Specification: Part of the patent application. It includes a detailed description of the invention, claims specifying the aspects of the invention for which protection is sought, and the extent of the protection being sought

Stability: Each generation of the variety should be similar.

Supplementary Protection Certificate (SPC): A special form of intellectual property (IP) that extends the protection of a patented active ingredient or combination of active ingredients present in a pharmaceutical or plant protection product after the expiry of the patent.

Tariff: A standard charge for use of copyrighted works. Usually applies to fees paid by users of musical works and cable companies for the rebroadcast of programs.

Trademark: A word, symbol or design, or combination of these, used to distinguish the goods or services of one party from those of another.

Trade Secret: Information about a product or process kept secret from competitors.

Trademark Agent: A a person whose name is entered on the list of trade-mark agents and who is therefore entitled to practice before the Trade-marks Office.

Trademarks Act: Central government legislation governing trade-mark registration in a country.

Trademarks Journal: A publication of the Trade-marks Office issued weekly and containing all approved applications and Office rulings.

Trademarks Office: The central government agency responsible for registering trade-marks in a country

Trademarks Records: The inventory of registered trade-marks and pending applications maintained by the Trade-marks Office.

Trademarks Register: The official listing of registered trade-marks.

Trade Name: The name under which a company conducts its business. Trade names may or may not be considered trade-marks, depending on the circumstances.

Traditional Knowledge [TK]: The knowledge, innovations and practices of indigenous people and local communities

Uniformity: All plants in a variety should look alike.

Universal Copyright Convention: A country that is party to the Universal Copyright Convention [adopted on September 6, 1952 in Geneva, Switzerland, or to that convention as revised in Paris, France, on July 24, 1971].

Variety: A genotype released for commercial cultivation either by State Variety Release Committee or by Central Variety Release Committee and notified by government.

WIPO (World Intellectual Property Organization): International body responsible for facilitating the granting of Intellectual Property Rights through out the world.

WTO Member Country: A Country that is a member of the World Trade Organization as defined in subsection 2(1) of the *World Trade Organization Agreement* Implementation Act.

World Trade Organization [WTO]: An international organization that regulates trade and tariffs between nations in order to ensure that trade flows smoothly, predictable and as freely as possible.

References

Alexander I., Poltorak and Paul, J. L. 2002. Essentials of Intellectual Property. John Wiley and Sons.

Alexander I., Poltorak and Paul, J. L. 2004. Essentials of Licensing Intellectual Property. John Wiley and Sons.

Amrita Narlikar 2005. The World Trade Organization. A Very Short Introduction. Oxford Uni. Press, USA.

Andrew, H. J., Phill, G. and Terri, R. G. 2006. Plant Diversity.

Arthur, R. M. and Michael, H. D. 2000. Intellectual Property: Patents, Trademarks and Copyrights [Nutshell Series]. West Group Publishing.

Bhagwati, J. 2005 In Defense of Globalization. Oxford Uni. Press, USA.

Carlotte Fiell and Peter Fiell 2000. Industrial Design A-Z. Taschen.

Carrie Russell 2004. Complete Copyright. An Everyday Guide for Librarians. American Library Association.

Clarisa, L. (ed) 2000. Intellectual Property Rights in Emerging Markets. Aei Pr.

David Pressman. 2005. Patent It Yourself. 11th edn. NOLO.

Donald,A Norman 2002. The Design of Everyday Things. Basic Books.

Gary, P. Sampson 2001. The Role of the World Trade Organization in Global Governance. UN Uni. Press.

Gert, W. R., Paul, V. K. and Bart, K. M. E. 2006. European Community Plant Variety Protection. Oxford Uni. Press.

Glenn Adamson 2003. Industrial Strength Design: How Brooks Stevens Shaped Your World ?. The MIT Press.

Henri Charmasson 2004. Patents, Copyrights and Trademarks for Dummies. For Dummies.

Henry, R. J. 2004. Plant Diversity and Evolution: Genotypic and Phenotypic Variation in Higher Plants. Southern Cross Uni., Australia.

Jessica Litman 2006. Digital Copyrights. Prometheus Books.

Jill Gilbert 2004. The Entrepreneur's Guide to Patents, Copyrights, Trademarks, Trade secrets and Licensing. Berkley Trade.

Joanna, R.J. 1997. Merchandising Intellectual Property Rights. John Wiley and Sons.

John Heskett 1985. Industrial Design: World of Art. Thames and Hudson.

Joseph, E. S 2006. Making Globalization Work. W. W. Norton.

Laurence R. Helfer 2005. Intellectual Property Rights in Plant Varieties: International Legal Regimes and Policy Options. FAO-UN.

Lee Wilson 1998. Trademark Guide. Allworth Press.

Lynn, H. and Cheryl, D. C. 2004. Design Secrets: Products 2: 50 Real Life Projects Uncovered. Rockport Publishers.

Mitsuo, M., Thomas, J. S. and Patros, C. M. 2006. The World Trade Organization: Law, Practice and Policy. Oxford Uni. Press, USA.

Patterson, L. R. and Stanley, W. L. 1991. The Nature of Copyright: A Law of Users Rights. Universal Georgia Press.

Paul Gouldstein 2004. Copyrights, Patents, Trademarks and related State Doctorines Foundation Press.

Peter Naray 2001. Russia and the World Trade Organization. Palgrave Macmillen.

Raymond Loewy 2000. Industrial Design. Overlook TP.

Richards, S. 2006. Patent, copyright and Trademark. An Intellectual Property Desk Reference. Amazon Book.

Richard, W. Stim 1999. Copyright Law. Thompson Delmar Learning 1st Edn.

David, J. Moser 2001. Music Copyright for the New Millenium. Artist pro. 1st edn.

Robert, A. G. and Jane, C. G. 2002. Copyright: Cases and Materials. Foundation Press, New York.

Robert C. Lind 2000. Copyright Law. Carolina Academic Press.

Robert, W. H. 2005. Intellectual Property Rights in Frontier Industries. Aei Pr.

Singh, Phundan 2016. Introduction to Intellectual Property Rights. Daya Publishing House, New Delhi.

Siva Vaidhyanathan 2003. Copyrights and Copy wrongs: The Rise of Intellectual Property and How it Threatens Creativity. New York Uni. Press.

Stephen Elias 2003. Trademark: Legal Care for your Business and Product Name. NOLO 6th edn.

Stephen Fishman 2001. Copyright Your Software. 3rd edn, NOLO.

Stephen Fishman 2006. Copyright Hand Book. What Every Writer Needs to Know? NOLO.

Books by the Same Author

A. Plant Breeding

1. Essentials of Plant Breeding
2. Plant Breeding: Molecular and New Approaches
3. Molecular Plant Breeding
4. Principles of Seed Technology
5. Seed Technology: At a Glance
6. Objective Seed Technology
7. Practical in Crop Breeding
8. Practical and Numerical in Plant Breeding
9. Numerical Problems in Plant Breeding and Genetics
10. Objective Science of Plant Breeding
11. Plant Breeding At a Glance
12. Plant Breeding [For Under Graduate Students]
13. Principles of Plant Breeding
14. Cotton Breeding
15. Heterosis Breeding in Cotton
16. Breeding Hybrid Cotton
17. Breeding Transgenic Bt. Cotton
18. Cotton Improvement in India
19. Glossary of Plant Breeding and Genetics
20. IPR and Plant Breeders' Rights [Subjective]
21. IPR and Plant Breeders' Rights [Objective]
22. IPR and Plant Breeders' Rights [At a Glance]
23. Introduction to Intellectual Property Rights

24. Intellectual Property Rights at a Glance
25. Intellectual Property Rights: Objective Analysis
26. Elements of Baby Corn
27. Breeding Crop Plants for Stress Resistance
28. Plant Breeding: Related Legislations

B. Genetics

29. Elements of Genetics
30. Genetics
31. Principles of Genetics [For UG Students]
32. Plant Genetics
33. Cotton Genetics
34. Objective Genetics
35. Genetics At a Glance
36. Objective Genetics and Plant Breeding
37. Objective Genetics and Plant Breeding [Hindi Edn]
38. Glossary cum Dictionary of Genetics and Plant Breeding
39. Molecular Genetics [Objective]
40. Molecular Genetics [At a Glance]
41. Molecular Genetics [Subjective]
42. Genetics and Man
43. Genetics Today

C. Quantitative Genetics

44. Biometrical Techniques in Plant Breeding
45. Application of Biometrical Techniques in Plant Breeding [Hindi Edn]
46. Objective Quantitative Genetics
47. Quantitative Genetics At a Glance
48. Quantitative Genetics

D. Biotechnology

49. Introduction to Biotechnology
50. Plant Biotechnology
51. Principles of Plant Biotechnology
52. Plant Biotechnology: At a Glance
53. Objective Plant Biotechnology

E. Competitive Examination, *etc.*

54. A Guide to Competitive Examinations of Agriculture
55. Objective General Agriculture
54. Glimpses of Cotton Breeding

Index

www.ingramcontent.com/pod-product-compliance
Ingram Content Group UK Ltd.
Pitfield, Milton Keynes, MK11 3LW, UK
UKHW020821290926
15470UKWH00003B/1290

9 789387 057685